PRAISE FOR *NEUROSCIENCE FOR COACHES*

"It is a rare book that changes the way one sees the world and how we all fit into it. The well-referenced knowledge in this book produces not only 'aha' moments for experiences in the past and present, but provides actionable insights for making coaching and managing easier in the future. Amy's style and explanations make a seemingly difficult topic not only accessible but a pleasure and exciting experience to learn. This book is essential for anyone who is serious about coaching and wants to go deeper into why it all works. Highly recommended."
Simon Powers, CEO and Founder, Adventures with Agile

"Every interaction with Amy Brann leaves you with 'aha' moments that you can't wait to put into practice. Her unique ability to combine deep neuroscience insights with engaging writing, relatable scenarios and actionable take-aways makes her work stand out."
Deepa Suryanarayana, Associate Director, Global Assurance Talent Team, EY

"Whether you're a coaching professional or a manager coaching your team, using neuroscience principles will help you create a more effective coaching experience. Amy Brann translates complex theories and ideas into accessible, practical and straightforward coaching strategies."
Gemma Leigh Roberts, Chartered Psychologist and Founder at The Resilience Edge

"One contribution of the Covid-19 pandemic has been to sensitize people to their bodies and the processes that enable them to think, feel, love, act and heal. Amy Brann has written a thorough Wikipedia of the brain in coaching. Her third edition is a powerful update of how our brain and neural processes affect us when interacting with and encouraging employees or clients to be open to learning and change.
Richard Boyatzis, Distinguished University Professor, Case Western Reserve University and co-author of *Primal Leadership* and *Helping People Change*

"Amy Brann's practical and pragmatic approach to neuroscience helps you to put insight into action. For coaches, this information can help you to improve your practice and increase the positive impact of your coaching conversations"
Helen Tupper, CEO, Amazing If and co-author of *The Squiggly Career*

"Neuroscience is a topic that no coach or manager can ignore. Amy Brann's well-researched book will provide readers with the latest theory and practical tools for their work. This book is comprehensive, clearly laid out and imaginatively written. I highly recommend Amy's work to those who wish to deploy the latest research in neuroscience to maximize people's performance, resilience, and well-being."
John Blakey, Founder, The Trusted Executive Foundation and CEO, executive coach and keynote speaker

"If you're a coach or leader, you need to read this book! Profoundly researched, practical, and easy to understand, *Neuroscience for Coaches* unlocks the secrets of how to work with your team more effectively for better results. Amy Brann captures the beautiful intricacies of the brain that will leave you armed with knowledge to use in every area of your life."
Marshall Goldsmith, Thinkers50 No 1 Executive Coach and *New York Times* bestselling author

"This book covers the thought-provoking area of neuroscience and its application to the field of coaching. Although potentially a challenging topic, it is written in an easy-to-read style. Amy Brann is to be congratulated on this third edition which many coaches will find useful and informative."
Stephen Palmer, Professor, Centre for Coaching, London

"Amy Brann possesses a wonderful ability to present neuroscience in understandable terms and, crucially, brings this alive with practical applications in the coaching relationship. Her zest for the topic runs through every page of this excellent book."
Keith Nelson, Director of Coaching Programmes, Moller Institute

"Amy Brann explains a difficult topic in plain, uncondescending language. Adding the relevance and potential usage of this new understanding, will add real value to coaches wishing to incorporate neuroscience into their practice. A highly recommended read."
Stephen Murphy, Global VP Thought Leadership & Development, EMCC

"In this work, Amy Brann does an exceptional job breaking down the science of our complex cognitive processes in a way that's easy to digest and apply to lead meaningful change. As the world's leading expert on Change Enthusiasm, I love reading about practical science that underpins what I teach. This is the stuff we all need to be aware of to take our organizations to the next level. Managers who put their people first, striving to nurture and inspire their teams, *must* read this book."
Cassandra Worthy, World's Leading Expert on Change Enthusiasm, keynote speaker and author

"When it comes to the field of neuroscience, Amy is simply the best. And I've looked! The way she is able to turn a scientific subject into something practical and meaningful in the world of work means that this book is a must read for anyone who intrinsically understands the value of neuroscience in the real, lived world, but hasn't quite been able to bridge the gap yet."
Hodl Whittaker, Learning Manager for Group

"An essential guide for coaches that want to understand how the brain works. Amy simplifies the science, demonstrates the coaching application, and importantly contextualizes why each concept is important in the workplace. An invaluable handbook for coaches, managers, leaders, or anyone that wants to get to know themselves better!"
Melissa J Sayer, Leadership Team Coach (Performance HUB and 6 Team Conditions) and adjunct Assistant Professor in Leadership & Coaching, Trinity Business School, Trinity College Dublin

Third Edition

Neuroscience for Coaches

How coaches and managers can use the latest insights to benefit clients and teams

Amy Brann

KoganPage

First published in Great Britain and the United States in 2015 by Kogan Page Limited
Second edition 2017
Third edtion 2022

2nd Floor, 45 Gee Street	8 W 38th Street, Suite 902	4737/23 Ansari Road
London EC1V 3RS	New York, NY 10018	Daryaganj
United Kingdom	USA	New Delhi 110002
www.koganpage.com		India

© Amy Brann, 2015, 2017, 2022

ISBNs
Hardback   978 1 3986 0421 6
Paperback   978 1 3986 0419 3
Ebook   978 1 3986 0420 9

British Library Cataloguing-in-Publication Data

A CIP record for this book is available from the British Library.

Library of Congress Control Number

2021953219

Typeset by Integra Software Services, Pondicherry
Print production managed by Jellyfish
Printed and bound by CPI Group (UK) Ltd, Croydon CR0 4YY

CONTENTS

PART TWO Brain chemicals

Discover the communication molecules of your mind 71

PART SEVEN   Neuroscience of not-so-classic coaching areas

ACKNOWLEDGEMENTS

As always with book writing, it is an accomplishment only achieved through the kindness, knowledge and skills of others. This time I'm particularly grateful to the following people.

My daughter Jessica for helping me improve my English skills. I'm more OK with fronted adverbials, prepositions and why commas must go where they do thanks to her. I'm also so grateful for the fun, laughter and playful spirit she brings to every day. She invites me to be a better coach and human being. Stuart, my husband, for his patience, love, and perseverance.

Jen from Synaptic Potential for a huge amount of research, without which this book wouldn't have been possible. And for her willingness to always learn and adapt. Fran for ongoing marketing energy and diving into whatever crazy new system we decide is sensible!

Jessie and Kate, two friends who are precious and so encouraging and who always make time to listen and support me. Rachel and Rob, the dynamic duo who provide such an inspiring example in life while offering the perfect blend of support and challenge.

PRELUDE

Introduction

We consider why neuroscience is becoming an expected area of knowledge for coaches and how to approach this fascinating field.

Part One – Brain areas

Here we look at various areas of the brain that are useful for coaches to be aware of. The chapters in this section are:

1 Prefrontal cortex

Affectionately thought of as the CEO or conductor of the brain responsible for higher-level cognitive functions, including attention and processing.

2 Basal ganglia

Key to storing routines, repetitive behaviours and thoughts – the home of habits.

3 Striatum

Involved in pleasure, reward, motivation, reinforcement learning, fear and impulsivity.

4 Insular cortex

Enables a degree of awareness and is involved in our ability to be 'in tune' with ourselves.

5 Amygdala

Part of the limbic system which is heavily involved in regulating our emotional life.

6 Anterior cingulate cortex

Primarily responsible for conflict or error detection.

7 Hypothalamus

Responsible for various metabolic processes and synthesizing and secreting neurohormones.

8 Hippocampus

Has roles in the consolidation of information from short- to long-term memory.

9 Cerebellum

Coordinates movements; controls posture, balance and fine motor movement; involved in motor learning.

The structure of each chapter addresses:

- What is this brain area?
- Why is it important to me as a coach/manager?
- What can I do with a client or colleague now having understood this?

Part Two – Brain chemicals

Part Two is about the brain chemicals that coaches will observe and hear about the effects of within their clients. The chapters in this section are:

10 Cortisol

Famously known as the 'stress hormone' because it is released in response to stress.

11 Dopamine

Involved in how we think and behave and our feelings of motivation, reward and attention.

12 Oxytocin

Involved in social behaviour, increasing trust, decreasing fear, increasing generosity and also cognitive functions.

13 Adrenaline

Best-known for its role in the fight-or-flight response.

14 Serotonin

Important for mood regulation, appetite, sleep, memory and learning.

15 GABA and glutamate

The king and queen of neurotransmitters.

The structure of each chapter addresses:

- What is this brain chemical?
- Why is it important to me as a coach/manager?

- What are some scenarios where knowing about this would enable me to best serve my client or colleague?
- What can I do with a client or colleague now having understood this?

Part Three – Foundational brain concepts

Part Three looks at the foundational things that a coach needs to understand about the brain and what this means to them. These underpinnings are vital for the next three parts of the book. The chapters in this section are:

16 Neurons and synapses

17 Neuroplasticity

18 Threat response

19 Neuroimaging

20 Working memory

21 HPA axis

22 Mirror neurons

The structure of each chapter addresses:

- What is this?
- Why is it important to me as a coach/manager?

Part Four – 22. Brain networks

The approach adds a level of understanding that will take you from a basic comprehension to an intermediate one. Rather than seeing each area of the brain as independent we look at three instances where multiple regions work together in a network.

Part Five – 23. The quantum brain

Part Five offers the opportunity to explore the possibility of viewing the brain through a quantum lens – hugely controversial and, subsequently, equally exciting.

Part Six – Neuroscience of classic coaching areas

This part looks at the neuroscience behind classic coaching tools, techniques and concepts that coaches will be familiar with. Neuroscience brings credibility, precision and a new perspective to these areas. The chapters in this section are:

Mindframes

Skills

Tools

The structure of each chapter addresses:

- What is this?
- Why is it important for coaches or managers?
- What is the underpinning neuroscience?
- What are any interesting studies in this area?
- How is it useful for managers who coach?
- What can I do with a client or colleague this afternoon?

Part Seven – Neuroscience of not-so-classic coaching areas

This part looks at some tools and concepts that are not frequently used by coaches or necessarily consciously addressed by them. The chapters in this section are:

36 Choice architecture

37 False memory

38 Trust

39 Fairness

40 Loneliness

The structure of each chapter addresses:

- What is this?
- Why is it important for coaches or managers?
- What is the underpinning neuroscience?
- What are any interesting studies in this area?
- How is it useful for managers who coach?
- What can I do with a client or colleague this afternoon?

LIST OF ABBREVIATIONS

ACC	anterior cingulate cortex
ADHD	attention deficit hyperactivity disorder
AON	action-observation network
CAT (or CT)	computerized axial tomography
CRH	corticotropin-releasing hormone
dACC	dorsal anterior cingulate cortex
DMN	default mode network
DMPFC	dorsomedial prefrontal cortex
dPFC	dorsal prefrontal cortex
dlPFC	dorsolateral prefrontal cortex
EEG	electroencephalography
EPR	Einstein, Podolsky and Rosen paradox
fMRI	functional MRI
FPN	frontal parietal network
GABA	gamma-aminobutyric acid
HPA	hypothalamic-pituitary-adrenal
HSAM	highly superior autobiographical memory
IA	interoceptive attention
LTP	long-term potentiation
MBI	mindfulness-based interventions
MEG	magnetoencephalography
MFN	medial frontal negativity
MRI	magnetic resonance imaging
MSG	monosodium glutamate
MTL	medial temporal lobe
NA	nucleus accumbens
NLP	neurolinguistic programming
OCD	obsessive compulsive disorder
PET	positron emission tomography
PFC	prefrontal cortex
PTSD	post-traumatic stress disorder
rIFG	right inferior frontal gyrus
SAD	seasonal affective disorder

SCT	social cognitive theory
SPECT	single photon emission computed tomography
SSRI	selective serotonin reuptake inhibitor
tDCS	transcranial direct current stimulation
vACC	ventral anterior cingulate cortex
vlPFC	ventrolateral prefrontal cortex
vPFC	ventral prefrontal cortex
vmPFC	ventromedial prefrontal cortex
VTA	ventral tegmental area
WOLF	work-related flow

Introduction

What is a coach?

'An expert in facilitating self-directed neuroplasticity.' Perhaps not the answer you normally hear, but certainly my favourite and one I use whenever I'm asked. Jeffrey Schwartz actually introduced me to this idea and I like to give credit to ideas I develop or share. Neuroplasticity is the brain's ability to change. People are incredible and have a huge amount of potential. Coaches are skilled at working with individuals to help them bring out the best in themselves and rewire their brains to make it easy for those changes to become the norm.

Going back in time, a *coche* was the word for a wagon or carriage. Essentially it was and still is a vehicle that takes a person from A to B. A human *coach* now appears in many fields. Perhaps the best-known is still the sports coach whose responsibility it is to intensively train the people they work with. In the last 30 years, however, the role and definition of a coach have developed. It is common for managers in organizations to act as coaches. It is even hoped that parents at times act as coaches to their children.

Inevitably you will have your own definition of what a coach is and what they are expected to do within that role. There are now a huge number of different models, strategies and training programmes around coaching. We believe that an understanding of neuroscience will become expected to *underpin* everything that sits on top of what a coach does. Neuroscience is not claiming to be better than anything you currently do, nor is the suggestion that this should replace what you do. Rather we suggest it will enhance your understanding of those you work with.

What is neuroscience?

Neuroscience is the scientific study of the nervous system. This discipline and its research are extensive and cover a lot, including the development of the brain, its anatomy and how it works. It looks at what happens when things go wrong, be it neurological, psychiatric or neurodevelopmental. It is a big field, and when positioning such a game-changing underpinning to coaching it is worth having a broad overview before focusing in on the parts most relevant to us. Today neuroscience has an interdisciplinary nature and liaises with disciplines such as mathematics, linguistics, engineering, computer science, chemistry, philosophy, physics, psychology and medicine. Several of these areas are of less interest to a coach.

Here are some of the areas you may hear or read about. *Affective neuroscience* is focused on how neurons behave in relation to emotions. *Behavioural neuroscience* is the study of the biological bases of behaviour, so it looks at how the brain affects behaviour. *Clinical neuroscience* is not usually of much interest to us as coaches because it focuses on disorders; however, a lot of interesting discoveries have come from this field. *Cognitive neuroscience* looks at the neural base for higher cognitive functions. *Cultural neuroscience* studies how our beliefs, values and practices are shaped by and shape the brain, minds and genes – a fascinating field that we hope will add lots to our understanding over the coming years. *Molecular neuroscience* looks at the individual molecules within the nervous system. *Neuroimaging* is the field that is expert in the various scanners that are used across the board. *Social neuroscience* brings the biological systems into the realm of social processes and behaviours. Both disciplines help refine and inform the other.

From the perspective of coaches, neuroscience is the field that can inform them on important things about the brain: things that are key to new ways that you work with clients and also that underpin things you are already familiar with. Neuroscience can explain why and how coaching works. It can enlighten coaches as to things to pay particular attention to. It can warn against other practices. As it is fundamentally concerned with the way that the brain works, its value to those working with other people's brains is quite vast.

At this point it is worth reiterating that neuroscience is not in any danger of taking over the skills, training and discipline that is coaching itself. It can certainly add to what is already there, but does not replace the great foundations that a coach has.

What can neuroscience offer the coaching world?

- An understanding of what happens when a person is being coached.
- Ways to create the best coaching environment for clients.

- For coaches who use particular models, an understanding of how they might work or an opportunity to stop using outdated concepts.
- A foundational understanding of the brain through which to evaluate everything.
- The opportunity to ask better-quality questions and consider what interventions would best serve your client.
- A focus on important considerations as to how to get desired action occurring: for example, mental stress leads to habit behaviour rather than goal-directed behaviour, so this informs how/when/what goal-directed behaviour should be attempted.

Our trained coaches tell us that it gives them a sense of credibility when they can explain the research that backs up what they are doing with their clients.

What is the vision?

The skills and knowledge that a good coach possesses can be so instrumental to a person living a good quality of life that it is my hope that in the future everyone is equipped. Imagine schools where teachers could effectively coach; organizations where managers, leaders and peers chose to coach; even, dare I suggest it, governments where coaching rather than attacking was the norm. With what we now know about the brain, and how people work, we have a huge opportunity.

People possess the precious ability to take in information, think and then act. This is grossly undervalued by individuals and by organizations. Coaching has the potential not only to help the person you are working with improve in these skills but to have a positive ripple effect.

Different ways of coaching

There are many models or ways of coaching out there. We've yet to come across one for whose practitioners we don't think neuroscience would be useful. Robert Dilts proposes that a coach has many roles at different times. He suggests that a coach may need to be in the role of guiding and caretaking, coaching, teaching, mentoring, sponsoring or awakening. Dilts elaborates on these roles in his excellent book *From Coach to Awakener* (2003). Here these roles are explored, sometimes adapted, and linked to various neuroscientific principles or areas of interest. The role of this guide or caretaker can be considered to be providing support with respect to the environment where the changes are taking place, like a guide who holds a hand to take a child or blind person down a path from one place to another. It is expected, of course, that the guide knows where they are going and how to reach that desired destination. The caretaker is tasked with providing a safe and supportive environment. Dilts suggests that in this role it is important to attend to the external environment, provide what is needed and remove distractions.

This role, like all of those set out by Dilts, will be working on multiple neurological levels, utilizing different brain areas and chemicals, and would benefit from paying attention to various things that the scientific research is pointing us towards. None of these do we know for sure without doing brain scans and taking blood samples, and even then there are things we would not know in each individual situation. So, when we talk about what is going on for people in their brains and bodies when they are being coached it is normally an informed guess. We do know that in a trusting relationship, such as between coach and client, the release of oxytocin is likely to be triggered. This will be beneficial for both parties. At a basic level, by supporting someone you might be lowering their cortisol levels and enabling them to hear more clearly. If there is a high level of trust, which, as an effective coach, you hope would come at some point if not initially, then any threat response may be decreased and the person may be able to think more effectively. By attending to the external environment and removing distractions, you are creating a better environment for the prefrontal cortex to focus and be effective and efficient in high-level information processing and decision making.

Small 'c' coaching can be considered as focusing on the behavioural level of performance. It is suggested that this resembles sports coaching and involves promoting conscious awareness of resources and abilities. Careful observation and feedback are utilized to strengthen a person's abilities. It is often context-dependent so it might be used in areas such as 'style coaching', 'business coaching' and 'wealth coaching', where the coach is suggesting new behaviours.

This behavioural level of focus is very important as it leads directly to our results. The question becomes: what can influence our behaviour? Here we are directed in neuroscience to the underpinning of habits, how we are motivated, how we make decisions and to interesting new areas such as nudging and priming.

Teaching involves new learning. It can involve increasing competencies or knowledge or strategies for thinking and doing. At times during a coaching relationship it may be important to be able to share something you think will help a person if you notice a gap in knowledge or competency. This is very obvious with neuroscience where there are things a person is doing that are not getting them the results they want, and where an acquisition of new information could really help. Learning new things and strengthening new competencies are important skills. Various brain areas are involved in these processes, for example the amygdala, hippocampus and basal ganglia.

Mentoring is described by Dilts as guiding a person to discover their unconscious competencies and overcoming internal resistance. He cites a belief in the person and validating their positive intentions as important components. The mentor has an impact on the beliefs and values of their mentee as their example unveils a little more of the person. Mentoring may involve mirror neurons or the action-observation network (covered in Chapter 22), especially when the mentor is a good example of something that the person being mentored wants to improve upon. It has been suggested that people learn tennis best by initially watching a good tennis player and then in essence copying them.

Sponsoring is described by Dilts as the process of recognizing and acknowledging the identity of the other person. This role of looking for and then helping to secure the innate potential that people possess is a critical role and one that is often forgotten in a corporate environment – with dire consequences. The idea is to connect with and validate someone deeply. This process enables the value within a person to manifest more fully. Sponsoring may be having effects at the level of the quantum brain. Things that would be more widely accepted now would be the potential of increasing an individual's serotonin and dopamine levels, activating the reward response and reducing any negative effects associated with loneliness.

The final role that Dilts proposes a coach holds is that of awakener. This is suggested to awaken others through the coach's own integrity and congruence. The idea here is to connect with missions and visions. The awakener role has an intensity about it, and is perhaps where real strengthening of new neural pathways occurs.

Why will a coach be expected to have an understanding of neuroscience?

It has been said that the gap between knowledge/theory and action is around 20 years. Since through the field of neuroscience we are now so much more equipped to understand what is going on in the brain it is only a matter of time before some people know more than others. First adopters in various spin-off fields utilizing neuroscience are enjoying the benefits of this fascinating research. Neuroeconomics is a great example of a field that was working well without the 'neuro' component. When economics turned to neuroscience to see what it could learn, additional fascinating insights emerged. This is an exciting time for the field of neuroeconomics and one we expect will become a routinely available module for those studying pure economics in the near future.

With the fast rate that we are learning about the brain and mind we have already amassed a lot of data. There is a proportion of coaches who are equipping themselves with this information. It makes sense, right? If you are going to be working with people's minds and brains then it is logical to know as much as you can about how these things actually work. You have started on this journey by reading this book. There are coaches who are also taking programmes in this area and equipping themselves with an in-depth understanding of neuroscience for coaches.

Organizations and individuals will continue to be discerning in how they choose their coaches. They will want the best, the most reliable and the most cutting edge. This means that coaches are expected to keep their knowledge up to date. When new information becomes available, especially knowledge that is so fundamental to the core of what one does as a coach, it is becoming expected that you are skilled in it.

How to approach the field

When coaches ask me this question I have several levels of response. The quickest answer is to approach the field *'curiously'*. If you only focus on one thing let it be your curiosity. This is a fascinating field that will develop a great deal more over the coming years, so now is the time to cover the basics and be curious about what you are exploring.

The next level of answer is to be *confident* in what you don't yet know. Whenever I give talks on this topic, or on topics that are core to our main work of applying neuroscience to organizations, people comment that it is refreshing to work with someone who is comfortable saying that they do not know something. Rest assured that the field of neuroscience is a huge one. It would take you a long time to get to know everything in the field, and new things are discovered every week. Experts in their fields tend to be just that, expert in their field. That field is often very specific. An expert in decision making may have very little awareness of the research being done on false memory. There is no shame in not knowing everything – however, there can be in pretending you do.

There has been suggestion that some people are turning to neuroscience in a superficial way, simply because it is a hot or sexy topic at the moment. That may be the case. People fall for gimmicks – they read a headline in a newspaper or even on what they consider a credible website and then use this surface-level piece of information to 'prove' something. Taking these shortcuts is very unwise in science. Evaluating the source, going back to the research and getting a full understanding is the best option to build reliable foundations and to be credible. In fields where you are working with people's brains and minds, such as coaching, there will always be a place for understanding the science. It is illogical to suggest that neuroscience for coaches could be anything other than a positive step towards a better future.

What are the boundaries/limitations?

For coaches who are very new to the idea of neuroscience, having a contribution to make to their field in this area can be a challenge. Their expectation is that there will be specific new jazzy tools and techniques that are 'neuroscientific'. They want to pull out of their bag their neuroscience tool and have their client go through a nicely packaged exercise. This desire is understandable: we like things to be simple, straightforward and easy to position with clients. However, good and experienced coaches know that there is a lot more to coaching than simply doing exercise after exercise with a client. This is where neuroscience comes in.

In this book we have at times included a section called 'What I can do with a client or colleague this afternoon?' Sometimes there is a tightly correlated exercise that is nice and easy to complete and that would be great to explore with a client, if this is how you choose to coach. At other times there is not. It is actually even more exciting when there is more to it than a simple exercise! (This is a mindset shift for some coaches, while obvious to many others.) Building up an understanding of how your client's brains and minds work is where the true magic lies.

Take, for example, the area of habits. We know a lot about how habits are formed in the brain and what areas are involved in these processes. There are many suggestions that can be used to support those you work with to gain increased control over their lives. However, there are also recent advances in our understanding of how habits work. We know that a small area of the prefrontal cortex is also involved in habits and specifically the 'turning off' of a habit. A natural question that people ask is: 'Can we just turn off bad habits, then?' Or, perhaps: 'How do we turn off that area of the brain, then?' Both of these questions are great and valid lines of thought – however, we don't have a safe and practical answer just yet. Instead, work with habits currently involves creating new ones that become stronger and more favourable than the old ones.

The challenge this book faces

One of the challenges for a book like this is that we are discussing a field that is rapidly developing. New studies are being done every week and some of these inevitably give us information that is contrary to previous data. This book is not pretending to offer you a snazzy new comprehensive 'solve all your coaching problems' model. It is a starting point: an introduction to neuroscience for coaches.

The opportunity once the basics have been covered is to deepen your understanding of how you apply this information, by going on a well-designed programme. Here you can connect with a supportive group of coaches who can quickly accelerate your practical application of this material. You will, of course, have your own coaching style and skill sets. Suggestions are offered throughout this book only to stimulate thoughts. You need supervized practice and reflection in order to integrate what you learn into your chosen way of coaching.

There are some classic examples of what we want to avoid. Imagine a senior leader of an organization who read a book on leadership – 20 years ago – and nothing since. There may be some key concepts that still hold true, but it is fair to say that they would be a far better leader if they stayed connected to the field of leadership and updated their knowledge and skills. There is a fear with writing a book of this nature that coaches think 'Well, that's it then, I know all I need to know about neuroscience'. That simply will not be the case. We can look at this from a couple of perspectives. First, the field will be continually evolving and so it will be beneficial to keep connected. Second, this book is not *it*: while the hope is that it will serve as a good introduction, really getting the information inside you and being able to use it confidently and fluently with those you work with is the next step.

How you use this information is in the hands of each individual coach. There is a danger with being introduced to new and exciting information that we want to rush in and share it with everyone. As coaches we need to remember that we are here to facilitate self-directed neuroplasticity. Our client's ability to think is a precious one – giving information inappropriately can destroy what Nancy Klein (2002) thoughtfully terms the 'Thinking Environment'. This precious environment deserves our respect.

What is the relationship between psychology and neuroscience?

In the past, and sometimes still in the present, there has been competition between the two disciplines. They have each, at times, looked down on the other. Psychology has accused neuroscience of being too reductionist; neuroscience has said that psychology can be inefficient and messy.

The benefit of psychology is that it can highlight phenomena that can prompt further exploration for mechanisms. Neuroscience can look at the more fundamental level and discover functional capacities of neural components. Links can be made here between behaviours and brain structures and function. Currently, while the two disciplines overlap, they are often working at different levels of things: neurons versus personality.

A useful analogy that can be drawn here is that of the Formula One car. Two main parties are involved in driving this car and, hopefully, winning races: one is the driver (the psychologist) – they are in peak physical condition; they know every detail of how their car performs (or are working towards this). If they want to go faster they know what buttons they need to press. When they have tricky bends coming up they know which way they need to turn the wheel and by how much. They have normally driven a track multiple times and can predict what the car will do at any given point. The other main party is the mechanics (the neuroscientists). These guys know what is under the hood of the car. They are focused on the nuts and bolts of the engine. They know that if they adjust some fluid by even just a small amount it will have an effect. They are interested in the most fundamental components of the car. Together the driver and the mechanics make a good team. Together they can each learn from one another and try different things.

A changed world

In the time since the second edition of this book was published and writing the third, the world has changed dramatically. We've experienced life as part of a world in the grips of a pandemic. It has changed individuals' day-to-day experiences and organizational ways of working, in ways that I don't think we will ever fully revert back from.

The message I am giving people who reach out for help, almost daily, is that now is the time to design back better. It is vital that we don't just 'build back stronger'. Before we can implement, we must learn and think and design. It is critical that we start with considering how we want life to be. What are the properties of a great life, for work, for individuals? Let's go back to basics in terms of how people can live to their fullest potential. Why? Because it is only then that they can give what they are truly capable of to their organization, to their team, to their partner, their kids, their friends and whatever they are choosing to do in life. If they are not good their contributions, in any sphere of life, will not be what it could be.

Who is this book for?

This edition is carefully written to benefit coaches who work independently with clients, and managers who work in all types of organizations. This is because over the last few years we (the team at Synaptic Potential) have been working much more closely with managers in various organizations across the world. They do such an important job. Managers are critical for unlocking the potential in individuals. We believe managers should act more like coaches. They should have the mindframes, skills and tools available to them to help their people be more effective and fulfilled. Coaches have always done an amazing job, and had a pivotal role in society in helping to develop people. Now we need managers to transition into helping more on a day-to-day basis too.

In an ideal situation to get the best out of individuals they would have both an amazing manager and a fantastic coach. Managers are with individuals physically or virtually on a regular basis. A coach is typically seen at most once a week. A manager is in the business and experiencing some of the same ups and downs. A coach can have a detached outside perspective. Both offer huge potential benefits.

What else has changed?

The science has moved on. Every year stacks of new studies are published. I'm grateful to my team who work hard to keep on top of the new developments to ensure we remain current, and ensure our clients continue to access the latest insights. There was a lot of new research to share in this book. In addition, the ways that organizations are open to using these insights has also moved forward. Never before have we worked with such informed, open and committed managers who are ready to make big changes internally.

The hope

My hope is that you will study these insights, go on to engage with our community who dig deeper, experiment and apply more fully and be shining examples of what is possible in the field of people development. It is more important than ever that we support people as society is reshaping. Let's base those reformations on what we know about how people work best.

High-Performing Neural Environment

What is it ?

A High-Performing Neural Environment (HPNE) is when your brain is in the most helpful state for whatever you want to do. The HPNE for creativity is different from doing focused work. Similarly, you'd regularly practise a different HPNE for building the character trait of compassion versus influencing by painting an attractive future. There are different brain networks, chemicals and patterns that are linked to different skills and characteristics. By understanding what the brain is up to during these different tasks, and how those states are influenced by different internal and external influences, you can trigger and maintain these different brain states necessary to fulfil potential in each thing one does. We're on a mission to help individuals and organizations recognize the science behind different HPNEs and how to create them. Throughout this book you'll read about how to create different HPNEs. As coaches your opportunity is to co-create one every time you coach someone.

Throughout this book we will share things top coaches have said to me in conversations. Here is a little background to the four coaches we will hear from.

Christian van Nieuwerburgh

Christian is a thoroughly nice guy. He is a well-respected executive coach, internationally recognized academic and sought-after consultant. He was an Associate Professor of Coaching at Henley Business School's Centre for Coaching and Behavioural Change and is now returning to the University of East London. Christian is particularly passionate about the power of coaching in educational settings. He is the editor of *Coaching in Education: Getting better results for students, educators and parents*, *Coaching in Professional Contexts* and *An Introduction to Coaching Skills: A practical guide*.

Kim Morgan

Kim is a much-loved coach. She created Barefoot, a coach training company offering externally certified University Master's Level Post-Graduate Awards in Coach Training in 2001. It also runs fabulous conferences. She offers fascinating insights as a trainer of new coaches, is an experienced coach herself and is adviser to organizations on creating coaching cultures. Kim is also the author of *The Coach's Casebook: Mastering the twelve traits that trap us*.

Marshall Goldsmith

Marshall Goldsmith has written many books including *Triggers: Creating behavior that lasts – Becoming the person you want to be* with Mark Reiter. The President of the World Bank, Jim Yong Kim, remarked: 'Marshall's coaching invites leaders to focus relentlessly on their behaviour. The leader's behaviour, as well as the team's behaviour, become the basis for great results and continuous improvement. This will be key to success for the connected, global, knowledge-driven companies of the future.' Marshall was named top executive coach by *Inc*. Magazine and the world's most influential leadership thinker by Thinkers50.

Susan Grandfield

After ten years in the corporate sector followed by another ten running her own training and coaching business, Susan began to recognize a need to help managers and leaders like herself deal with the challenges, the uncertainty, the pace and the demands of work and life. She attended the first Neuroscience for Coaches programme and went on to develop her combined approach based on mindfulness and neuroscience to create a highly successful way of helping leaders, managers and their teams to 'press Pause' and regain a sense of balance in what they do. Susan introduces her approach through 1:1 coaching, group workshops, at speaking events and through her writing. Her most recent publication is *Pressing Pause: A practical guide to mindfulness in everyday life* (you can download it at www.sgdevelopmentsolutions.com).

PART ONE
Brain areas

'What part of the brain does that?' This common question is natural, but often unanswerable, at least accurately! With all the advancements in our ability to see into the brain while it is doing things, particularly through MRI (magnetic resonance imaging) scans and EEG (electroencephalography), people frequently have the expectation that we can explain everything by linking it back to a specific area of the brain 'lighting up'. Fortunately, the brain is a little more complex than this. Despite possessing a fixed anatomy our brains exhibit diverse functionality. The question of how our brain does that is explored further in Part Four on dynamic functional connectivity. However, we first need to start with the foundations and then build up from there.

One of the most wonderful realizations felt by many of the participants of our Neuroscience for Coaches programme during the first module is that things are more layered than they previously realized. There is more going on in their clients' brains than they thought. Initially some coaches and managers hope that they would be able to identify, with certainty, an area of the brain that is responsible for a certain problem their client is dealing with – and then fix it. They quickly realize, however, that this will not be the case.

So what is the point of learning about the different brain areas? There are several answers to this question. The first is that our knowledge shapes our filters and our questions. When helping to unlock potential in people, these are two of the most valuable things we have. Consider your filters and questions before you had any coach training... and afterwards. One hopes that they are quite different. It is the same once you learn about the different areas of the brain (and everything else you cover in this book and then during your further studies).

The second point of importance for learning about the different brain areas is to give you a solid foundation. The concepts we look at in this book have clear and direct applications to coaching, so when we start talking about mindfulness and the effect it has on the prefrontal cortex then it makes sense to know what the prefrontal cortex is, what it looks like when it isn't working well, what it needs to work well – and lots more.

The third benefit is that at times you may want to discuss the areas of the brain with your clients or colleagues, offering them an additional perspective to their situation. We cannot know for sure whether a particular brain area is involved with a problem or situation they are working on, but most of the time we don't need to. There are thousands of examples of how this would work in real coaching situations; in Chapter 2 on the basal ganglia we will look at just one. For a client who is working on changing their habits, being able to explain how and why the brain works with habits can be really useful to them.

It is important to bear in mind as we go through Part One that each time we get to the section 'What can I do with a client or colleague now having understood this?' what is being offered are only suggestions. What will work and be appropriate in one situation is not the same for another. You are best placed to blend the science with your art at this point and balance everything you know in order to make an informed decision about what would best serve the person you are working with.

Warning

We start with the basics for good reason. Students who work with us often find the first module, where we teach the anatomy and biochemistry, the hardest. They go on to tell us that the firm foundation was very rewarding as they progressed into the more practical insights from neuroscience. We've even had lecturers, trainers and L&D professionals say that they have re-structured their programmes as a result of understanding more deeply how we learn.

There are now many other books and even programmes out there which cherry-pick the neuroscience they want to focus on. This approach is often what credible scientists and other professionals object to. Overall it has tainted people's feelings towards using insights from neuroscience. The surface-level approach means big chunks of important information are missed. By choosing quick and easy you often sacrifice correct and useful.

Before we dive into the different anatomical areas of the brain it is worth taking an overview perspective and touching on something you may have already heard of: the triune brain.

What is the triune brain?

The triune brain is a model that was developed by physician and neuroscientist Paul D MacLean in the 1960s (MacLean, 1990). It describes the evolution of the brain.

The hypothesis is that there are three parts to the brain: the reptilian, the paleomammalian (better known as the limbic) and the neomammalian. The first, the reptilian, comprised the basal ganglia and it was suggested that this complex was responsible for the instinctual behaviours. So drivers like aggression and dominance were said to come from this reptilian brain. It is the oldest in evolutionary terms.

The next in evolutionary terms is the paleomammalian complex. It contains the brain areas including the amygdalae, hypothalamus, hippocampal complex and cingulate cortex. The 'limbic system' was used to refer to this group of structures that are so interconnected. MacLean suggested that this limbic system was responsible for things like motivation and emotion involved in feeding, reproductive behaviour and parental behaviour.

The newest part of the brain, according to MacLean, is the neomammalian complex. This is basically the cerebral neocortex. It is responsible for our higher cognitive functions such as planning, perceiving and language. While the model is still in favour in some circles there are some challenges. Some of the arguments against this model are that:

- Some recent behavioural studies call into question MacLean's thoughts around the reptilian brain.

- Anatomically, the basal ganglia were also found in amphibians and fish (not just reptiles).

- Birds have been documented to have sophisticated cognitive abilities and language-like abilities.

- Not only paleomammals care for their children/offspring – even some fish do it.

- Crucially, it is now suggested that the neocortex was present in the earliest mammals.

Linked to this topic is the concept of the limbic system itself. There is a lot of controversy around this term. While some still use it freely, others say there is too much wrong with it. For example, the hippocampus is a key part of the limbic system. The idea is that the limbic system is the emotional centre of the brain while the neocortex is the cognitive zone. When damage is done to the hippocampus we see that people suffer from memory problems. Neuroscientists keep changing the boundaries of the limbic system. Joseph LeDoux suggests we should abandon it altogether.

Why is the triune brain model important to me as a coach/manager?

This model is fairly widely espoused in non-scientific circles. Having an understanding of it and what the science community are saying about it puts you in a more informed position.

Yet while this model is very attractive in its simplicity, it doesn't have the backing of neuroscientists across the board. So the suggestion here is to use with caution.

Prefrontal cortex

Affectionately thought of as the CEO or conductor of the brain responsible for a lot of higher-level cognitive functions, including attention and processing.

Prefrontal cortex

What is this brain area?

The prefrontal cortex (PFC) is an area of the brain found at the front of the head, behind your forehead in your frontal lobes. It is made up of lots of different subregions relating to specific cognitive control operations. While we affectionately refer to it as a central executive… this isn't supported neurally. It is suggested that the PFC supports multiple cognitive-control systems that may collectively be conceived of as a distributed executive (Nyberg, 2018). Evolutionarily it is one of the newest brain areas and takes the longest to mature. The PFC journey of maturation parallels the maturation from childhood through adolescence into adulthood.

The dorsal prefrontal cortex (dPFC) is interconnected with brain areas that are involved with attention, cognition and action. The dorsolateral prefrontal cortex (dlPFC) is known for being involved with short-term memory and is implicated in self-control. The ventral prefrontal cortex (vPFC) is interconnected with brain areas involved in emotion. The ventromedial prefrontal cortex (vmPFC) performs risk benefit analyses after receiving inputs from the amygdala and other parts of the frontal lobes.

There is a famous story in the neuroscience world that involves a man called Phineas Gage. In 1848 Gage was a 25-year-old construction worker. He had been a respected friend to the men he worked with and was good at his job. One day there was an explosion while Gage was tamping powder with a fuse in a hole (before sand had been poured in). The rod weighed almost a stone, was 3 feet 7 inches in length and 1.25 inches in diameter. As it flew into the air it pierced Gage's left cheek, going through the base of the skull, traversing the front of his brain and exiting at high speed out of the top of his head. The rod mostly destroyed his frontal lobe, including his prefrontal cortex.

The first thing that amazed bystanders was that Gage was able to walk, talk and be 'normal' (remember that part of his brain was now lying some distance away on the end of a rod). The second thing is that he survived the predictable infections that in 1848 were treated without antibiotics. Yet although he was physically intact, his previous likeable personality was not. He became 'fitful, irreverent, indulging at times in the grossest profanity which was not previously his custom, manifesting but little deference for his fellows, impatient of restraint or advice when it conflicts with his desires, at times pertinacious obstinate, yet capricious and vacillating, devising many plans of future operation, which are no sooner arranged than they are abandoned... A child in his intellectual capacity and manifestations, he has the animal passions of a strong man.' The damage to his brain – as we can see (thanks to the skull being dug up, examined, and scanned) – was in the prefrontal cortices. His personality was changed forever and he was unable to make good choices.

The PFC is connected to many other regions of the brain and body. The distributed network of connectivity with other regions, such as the hippocampus, amygdala, striatum and PCC is essential for modulating several aspects of executive function (Girotti, 2018). There are important links to regions involved in social interactions (the nucleus accumbens, amygdala, ventral tegmental area, hypothalamus) (Bicks, 2015). The lPFC's ability to implement goals relies on reciprocal interactions with a wide range of distributed regions – striatum, affect; anterior insula, mental states; PCC, as well as the regions that initial goals through conflict detection and goal selection mechanisms; ACC (Carlson, 2018). The vmPFC reward processing and decision-making functions are thought to depend, in part, on interactions with the ventral striatum and amygdala (Hiser, 2018).

More connections include the mPFC having extensive downstream projections to regions from the amygdala to the brainstem. This provides a substrate for downstream regulation of autonomic and neuroendocrine balance with influences on parasympathetic and HPA activity (McEwen, 2013). We know that the PFC is involved in mental health challenges. It is implicated in nearly all affective disorders, including depression (Dixon, 2017).

Why is it important to me as a coach/ manager?

The PFC is a highly significant area of the brain for anyone managing or coaching people to be aware of.

The PFC is involved in:

- all our executive functions;
- our ability to plan;
- decision making;
- expressing our personality;
- aligning our thoughts and actions with internal goals;
- moderating social behaviour.

The psychological term 'executive functions' covers a wide range of activities. It includes the ability to determine good and bad, same and different, processing the future consequences of current activities, working towards defined goals, suppressing socially undesirable urges and predicting outcomes.

A professor of mathematical psychology, David Meyer, performed a wonderful experiment. He invited a group of young adults to test what happens when people are switching between things quickly. The experiment involved the participants working out mathematical problems and identifying shapes. When they had to switch between the tasks their accuracy and speed decreased as compared to when they could perform one task and then the other. In some cases, multitasking added 50 per cent to the time required. Imagine a person working a 12-hour day and achieving the same results as a person working 8 hours, but with more mistakes and less elegance simply because the 12-hour day person was multitasking. 'Not only the speed of performance, the accuracy of performance, but what I call the fluency of performance, the gracefulness of their performance, was negatively influenced by the overload of multitasking' (Meyer et al, 2001).

Under stress our prefrontal cortex does not function well. Consider the implications of this reality! The neural circuitry and neurochemistry of the PFC can be changed by our experiences. Stress during childhood and adolescence can be particularly impactful on the structure and function of the PFC, although these effects are not necessarily permanent. In adults it has been shown that even mild acute (as opposed to chronic) stress can have a rapid and dramatic impact on our PFC's ability to function. It can affect creativity, flexible problem solving, working memory and other processes. Many stressed-out people find it quite a wake-up call to discover that chronic stress literally can cause the architecture of their PFC to change. One study looked at the effects of psychosocial stress after one month. It showed that the subjects had impaired attention control and disrupted functional connectivity within a frontoparietal network that mediates attention shifts. The good news was that after one month of reduced stress their responses became normal again. While it has been great to work with so many organizations now taking mental health more seriously during the Covid-19 pandemic, the deeper understanding of what stress does to people needs to be understood.

In a scenario where someone feels overwhelmed, with too much work to do or with a goal that is too big, then it can be useful to break it down into what needs to happen today. For example, if a budget meeting takes place and results in a huge financial goal that needs to be met, it could help to break it down into a weekly target. This could re-engage the dorsolateral PFC (short-term memory) by chunking things down into small short-term bite-sized pieces. Conversely, when someone is feeling very anxious it may be that their ventromedial PFC is activated. Consider a personal coaching scenario where a person's spouse tells them that they no longer feel loved by them. The person may feel an immediate threat, imagining what would happen if they broke up and how difficult things would be in lots of ways. An aim here would be to reduce the immediate threat. Change the time frame from a very short-term one to a long-term perspective. It also may be useful to shift from an empathetic state to a cognitive one. Get planning – what could be done over the next month to help move the situation in the direction your client wants it to go? The same principles apply for a professional scenario where a manager is supporting someone who is nervous about their performance review (if you still do the old style of these).

The PFC is far more involved in our emotions than was previously realized. It has a critical role in assigning value. It's key to interpreting exteroceptive sensations (the sensory information from the external environment) based on context and goals. These representations that are created are important in contrasting the desirability of different outcomes (Dixon, 2017). From the perspective of the manager and coach it is helpful to think of emotions as being constructed. They can be influenced and even trained over time.

Making a High-Performing Neural Environment (HPNE) possible requires managing the PFC. When it is in peak condition you have far higher chances of the whole brain working well. This is critical for:

- temporal organization of behaviour;
- representing and producing new forms of goal-directed actions (Carlen, 2017);
- regulating emotions (Dixon, 2017);
- teamwork involving pro-social behaviours (Carlson, 2018);
- top-down control over paying attention (Paneri et al, 2017).

What can I do with a client or colleague now having understood this?

In some coaching and managing situations it might be appropriate to step into the teacher role and share some fundamental things about the PFC.

This would be especially relevant to anyone working on their productivity, their efficiency or effectiveness.

What do you need to know about the prefrontal cortex?

- Its role is classically described as like that of a CEO (if you are business minded) or a conductor (if you are musically minded) – in short, it is the boss, it is responsible for your 'executive' functions, which means your ability to think, choose, plan, etc.
- Over the years it has developed considerably, and recent studies indicate that meditation increases its size further still (see Chapter 27 on mindfulness).
- It is hugely energy-hungry, but gets drained quickly.
- Stress impairs its ability to use energy.
- Structurally, it is part of your frontal lobe, which is the area at the frontmost part of your brain.

So what is the motivation for ensuring our PFC is in good working condition? When our PFC is not working optimally we find ourselves:

- feeling lazy;
- feeling lethargic;
- uninspired;
- easily distracted;
- being poor at completing things;
- fixing attention on repetitive negative thoughts;
- being disorganized;
- being forgetful;
- feeling overly emotional.

On the other hand, when the PFC is working in tip-top condition you can look forward to:

- intentional awareness;
- a long attention span;

- being able to contemplate possibilities;
- being able to plan;
- being able to stick to the plan;
- focusing easily.

When the PFC is overworked it doesn't function well at all. To rectify this situation, check out the recommendations below. If the PFC is not functioning properly this makes it very difficult to be effective, let alone efficient, which is often very frustrating when we are trying to get through the day. At such times it is quite common to feel that something is wrong and to perhaps fall back on old habits. This is a form of survival mechanism. When people find themselves micromanaging, controlling or punishing there is often a brain deficit involved. This doesn't always mean that something is overactive; an underactive PFC causes problems too. Dopamine, you'll soon find out, is the neurotransmitter that has many functions for the brain including reward, motivation, working memory and attention. When the brain has a lowered ability to use or to access dopamine it means that other brain areas aren't being quietened down in order to enable the brain to focus on one thing at a time. This makes life difficult for us because we struggle to focus, so become less efficient.

As a coach or manager you could also step into a mentoring role and lead by example, by following these recommendations:

- Turn off the e-mail function on your mobile phone in the evening so that your brain has downtime before you start work the next morning.
- Prioritize the big weekly tasks first then smaller tasks on a daily basis (experiment with the night before or first thing in the morning to see what works best for you).
- Turn on e-mails only at certain times of the day.
- Monotask for short- and long-term benefits.
- Pick small things you know you can achieve and then do them – this could raise your dopamine levels.

Equally, you could work with your people to help them experience their prefrontal cortex working well. Take them out of their normal working and thinking environment – for example, if they are normally office bound, take them into the countryside. Help them to connect with their senses, feeling the wind in their hair, the sun (hopefully) on their skin and breathing in all the different smells. The hope would be that this change of environment would invite some fresh thinking.

Basal ganglia 02

Key to storing routines, repetitive behaviours and thoughts – the home of habits.

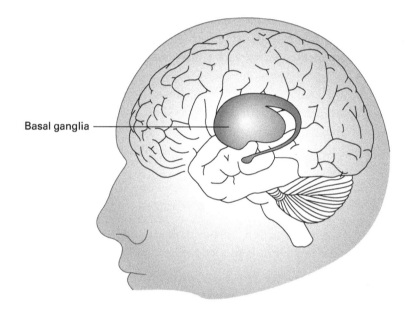

Basal ganglia

What is this brain area?

A more accurate question is 'what are they?' There are billions of basal ganglia and they are key to storing routines, repetitive behaviours and thoughts. The subcortical nuclei that process information received from all areas of the cerebral cortex send this data back to the frontal cortex and other brainstem structures involved in the control of motor and emotional responses (Dacunha, 2012). Various studies have shown that the inputs and outputs of each basal ganglia structure are topographically organized, which suggests it consists of separate circuits that have distinct functions (Kim, 2015). The main components of the basal ganglia – you do not need to remember this, it is included only because some people enjoy a high level of detail... I rarely refer to these components in practice – comprise the striatum (caudate nucleus and putamen), the globus pallidus, the substantia nigra, the nucleus accumbens and the subthalamic nucleus. You will hear lots more about the striatum and the nucleus accumbens in the following chapter and these will start to stick in your brain with increased exposure. The patterns that your daily life is filled with are recognized, stored and repeated by your basal ganglia. Along with habits, the basal ganglia are associated with cognitive and emotional functions.

The BG works using 'if-then' coding and has lots of mini programmes, or maps, stored. Every habitual thing you do has a map. For example, if when you turn on your phone in the morning you immediately check LinkedIn to see what is happening, then your basal ganglia is coding 'if the phone is turned on in the morning then check LinkedIn'. The instructions are present for exactly how to do this. This brain region is very well connected. Information from most other brain areas is sent here.

A study that looked at the recognition of patterns in the basal ganglia when the subject is unconsciously aware of a pattern being present was carried out in Montreal. The volunteers were split into two groups. Both groups had to press one of four buttons on a keyboard to correspond with a light flashing in a certain location on a screen. One group watched the light appear in a random order while the other group were given a pattern. This pattern was difficult to consciously detect most of the time, but their basal ganglia could be seen via the brain scanner to be picking it up. It has been said that it can take as few as three repetitions for the basal ganglia to pick up a pattern. At times, individuals were able to identify the pattern and they could type their sequence 30–50 per cent faster than if there wasn't a pattern in the sequence. This underpins how a large proportion of our day is run on autopilot.

The BG are influenced by signals coming from other parts of the brain. The main basal ganglia input nucleus is the striatum which receives topographical excitatory projections from almost the entire cortical area (Florio, 2018). This is important to be aware of for a fuller understanding of the brain because there is often a balance of inputs occurring. A classic example are emotional and rational inputs into a decision-making process. When we switch from one behaviour to another, which involves the activation of the motor system, the signals from other parts of the brain are key. For example, the prefrontal cortex may convey a signal telling the basal ganglia to enable a specific action to occur. Therefore, understanding how the prefrontal cortex works, and the other brain areas, is also important to fully understand how the basal ganglia will respond. We know the BG is involved in various executive functions including planning, working memory, attention, problem solving, verbal reasoning, behavioural inhibition, mental flexibility, multi-tasking and initiation and monitoring of action (Dacunha, 2012). The ability to maintain focus in the execution of specific behaviours and to adapt appropriately to external and internal cues are key deficits observed in diseases that affect the BG (Haber, 2014).

The neurotransmitter dopamine plays an important role in the basal ganglia. (It isn't the only chemical though, also serotonin, acetylcholine and norepinephrine (Kim, 2015). A neurotransmitter is a chemical that communicates a signal across the gap (called a synapse) between two neurons. There is a lot of evidence that the BG have a role in motivation, especially the limbic part of the BG, the nucleus accumbens (NA), ventral pallidum and ventral tegmental area (VTA). The dopamine neuron projections from the VTA to the NA is important in the reward system of the brain. Things that we find primarily rewarding in life such as food, sex and addictive drugs all trigger the VTA dopamine system.

Why is it important to me as a coach/ manager?

The BG are fundamental to habits and to general automatic responses. People you work with are likely to be experiencing a large part of their day on autopilot. Many of the thoughts they have will be the same as yesterday's thoughts. Most of the actions they take will be similar to yesterday's actions. The interactions they have with family and colleagues are likely to be quite repetitive. This is all desirable from the efficient brain's perspective. However, when the content of these repetitions is not getting the results the person wants, then things need to change. It can be useful for clients to understand how their brain is wired and how they can change that wiring. There is great benefits to our basal ganglia working in the way that it does and empowering clients to know how to work with their brains rather than against them. We want our brains to use precious energy for processing novel information and to think, but for this to be the case we need habits that support us. We explore these in depth in Chapter 35 on habits.

The BC are also critical in enabling people to achieve goals (Hikosaka, 2017). They are useful in cognitive planning particularly of purposeful motor activities (Florio, 2018). There are circuits that are involved in skills that enable people to adapt to both volatile and stable environments (Kim, 2015). When we need to make decisions, different BG areas work together to coordinate behaviour and what is needed at that moment (Florio, 2018). What we choose to do is broadly considered to be motivated by what is coded to be rewarding to us. Goal directed behaviour is vital for survival. Managers are uniquely positioned to help people adjust how different actions are perceived and therefore how much effort is behind them.

What can I do with a client or colleague now having understood this?

As previously acknowledged, coaching styles vary widely and many of these different approaches lead to the facilitation of self-directed neuroplasticity. One thing you could consider doing with clients or colleagues involves the sharing of some of this information. Perhaps at a time when you are focusing on habits you could introduce the basal ganglia – by sharing the positive intention of this part of the brain, and how we rely on maps and circuits for things so that we do not have to think about them every time we do them. You should also share the potential downside to this, being that when we want to do something differently it can be hard. Our brain wants to be efficient and free up vital resources to be used for things that are new and require attention. Yet by seeing habits as things that serve us, we can start to develop some great foundations.

Considering next coaching at a very practical level, exploring a person's 'if-then' codes can be very illuminating. It may be that a whole session is required on addressing the 'if-then' coding that is and isn't working for them. Let us take the example of someone working on a weight-loss goal.

Remember how powerful holding a mirror up to someone can be (when you're coming from a good place). If they see a piece of chocolate cake and then they eat it this may not be ideal. If they get home from work and then see the sofa and fall exhaustedly onto it, rather than turning around and heading out to the gym, then changes also may be a good move. If they say to themselves regularly 'I deserve a treat' and then have an Indian takeaway, that may not be ideal. At work if people are asked to do something and they always say yes this can cause problems. If they're told about changes coming up and they start to worry that can erode resilience. These are all if-then plans coded by the basal ganglia. Simply by talking about these if-then plans and verbalizing more appropriate ones means it is more likely that the latter will be enacted. How long it would take to make new habits depends on a lot of variables, so I prefer to focus on putting the building blocks in place that will give people the best chance of turning to these new habits when they need them rather than to say 'in 30 days you'll have broken old habits and replaced them with new ones' (as this just is not necessarily the case).

In a more corporate environment our attention is drawn to the role of dopamine in the basal ganglia. Dopamine helps us to feel motivated, which is obviously very important at work. Dopamine is involved in influencing reinforcement learning processes by conveying a reward prediction error (Caligiore, 2019). Lots of energy, time and money gets invested by organizations into trying to motivate their employees. As a manager in the role of a coach or a leader you have an opportunity to educate or mentor those you work with. Synaptic Potential have worked with many organizations to help them transition from seeking to motivate to engage employees. From the perspective of the brain this has many benefits. Check out 'Engaged' by Amy Brann for more on this topic.

At a very basic level, living a healthy lifestyle is important for brain health as well as the health of every other organ. People could ensure that they are getting enough dietary tyrosine, which is a basic building block of dopamine and is found in avocados, bananas, meat, sesame seeds and almonds. Other foundational things to consider, and whether or not these are in place, are people eating enough antioxidants from green and colourful vegetables and fruits. Antioxidants can reduce free radical damage to brain cells, including the cells that produce dopamine. None of these dietary factors will 'boost' your dopamine; healthy diets are an important foundation though.

If you work in an organization, could your team do a healthy living month? This could, using some gamification principles and behaviour-change concepts, improve the choices people are making and ultimately lead to a healthier, more robust brain. Experientially helping people to understand how their basal ganglia work is another option that is open to you. Gamification is a relatively new process that is capitalized on by technology companies. Some principles that could be mapped across into organizations – this would need testing to see what concepts work best – include:

- Create a public 'scoreboard' for people to mark on it their intended workouts for that week and their actual workouts, and have stars given each week to people who do what they said they would do.

- Create a 'buddy system' for people to hold one other individual accountable to their daily number of steps targets.

- Have a bowl of fruit in the office with a set of smiley-face stickers next to it, then each time a team member eats a piece of fruit they should put a sticker next to their name on their personal scorecard – their aim is to get to two per day within four weeks.

Finally, since the basal ganglia are influenced by the prefrontal cortex (PFC) it is worth teaching your people about how the PFC functions well – in order to help them get the best from themselves overall.

A study that looked at the recognition of patterns in the basal ganglia when the subject is unconsciously aware of a pattern being present was carried out in Montreal. The volunteers were split into two groups. Both groups had to press one of four buttons on a keyboard to correspond with a light flashing in a certain location on a screen. One group watched the light appear in a random order while the other group were given a pattern. This pattern was difficult to consciously detect most of the time, but their basal ganglia could be seen via the brain scanner to be picking it up. It has been said that it can take as few as three repetitions for the basal ganglia to pick up a pattern. At times, individuals were able to identify the pattern and they could type their sequence 30–50 per cent faster than if there wasn't a pattern in the sequence. This underpins how a large proportion of our day is run on autopilot.

The basal ganglia are influenced by signals coming from other parts of the brain. This is important to be aware of for a fuller understanding of the brain because there is often a balance of inputs occurring. A classic example is emotional and rational inputs into a decision-making process. When we switch from one behaviour to another, which involves the activation of the motor system, the signals from other parts of the brain are key. For example, the prefrontal cortex may convey a signal telling the basal ganglia to enable a specific action to occur. Therefore, understanding how the prefrontal cortex works, and the other brain areas, is also important to fully understand how the basal ganglia will respond.

The neurotransmitter dopamine plays an important role in the basal ganglia. A neurotransmitter is a chemical that communicates a signal across the gap (called a synapse) between two neurons. There is a lot of evidence that the basal ganglia have a role in motivation, especially the limbic part of the basal ganglia, the nucleus accumbens (NA), ventral pallidum and ventral tegmental area (VTA). The dopamine neuron projections from the VTA to the NA is important in the reward system of the brain. Things that we find primarily rewarding in life such as food, sex and addictive drugs all trigger the VTA dopamine system.

Striatum and nucleus accumbens

Involved in pleasure, reward, motivation, reinforcement learning, fear and impulsivity.

Striatum

Nucleus accumbens

What is this brain area?

The striatum is the core component of the basal ganglia. The striatum receives input from the cerebral cortex and gives input to the basal ganglia, so, it acts as a sort of relay station. Why cover it separately from the basal ganglia? Because there is so much to cover, and a lot of research mentions the specific components.

Although at this stage of the book there are a lot of anatomical terms being introduced, they will become more familiar as you read them again in other contexts! The ventral striatum is the part of the striatum closest to the face and consists of the nucleus accumbens and the olfactory tubercle. The dorsal striatum is the largest part of the basal ganglia and is positioned towards the back of the head (ventral always means towards the front and dorsal towards the back of the head) – it includes the caudate nucleus and putamen.

The nucleus accumbens is part of the ventral striatum, which you'll remember is part of the basal ganglia. As is often the case we have two, one in each hemisphere. Technically you can divide the nucleus accumbens into the core and the shell, with different functions, but we'll refer only to the nucleus accumbens (NA) as a whole. We know that the NA is involved in pleasure, reward, motivation, reinforcement learning, fear, addiction, impulsivity and the placebo effect.

The NA is well known for its role in addiction (Harris, 2019). Functional imaging studies show that environmental cues that are linked to people taking addictive drugs trigger the release of dopamine in the NA. For most coaches and managers, what is of more interest is its role in processing rewards and novelty in memory encoding. It is activated when we see pleasant, emotionally arousing pictures and during visualizations of similar such scenes. It is also involved in the regulation of emotions induced by music.

One of the old but classic studies was performed in the 1950s (Olds, 1956). Electrodes were put into the septal area of a rat and the rat could push a lever to stimulate the electrode, by so doing stimulating that area of its brain. They found that the rat would continue to press the lever, preferring it to even stopping to eat or drink. The conclusion was drawn that the area is the 'pleasure centre' of the brain and is involved in reinforcement learning. In human experiments the level of dopamine in people's NAs went up when they believed they were receiving money, for example, or when heterosexual men were shown pictures of attractive women. The NA is also important in the feeling of 'wanting'. We get a flood of dopamine in the NA when we are expecting or anticipating a reward. This facilitates lots of different approach-oriented and goal-oriented behaviours. It has been said the NA helps an organism get closer to the good things in life as well as away from the bad things! (Floresco, 2014).

Axmacher et al (2010) looked at how we process unexpected events. It was known that we are more likely to remember unpredicted events than the predictable and routine ones. The relationship between this novelty processing and the hippocampus and NA has been explored. Prior to this study it was known that the hippocampus is involved in memory formation and the NA is involved in processing rewards and novel information. It was thought that the information transfer between the two brain areas were greater for unexpected events.

The study was an unusual one and the data collected had some limitations but the researchers found that the unexpected stimuli enhanced an early and late electrical potential in the hippocampus. The late signal was linked to the memory for the unexpected stimulus. In the NA there was only a late signal and it was larger for the unexpected stimulus. It appears that the two brain areas work as a team, as stated by the study leader Dr Axmacher: 'Our findings support the idea that hippocampal activity may initially signal the occurrence of an unexpected event and that the nucleus accumbens may influence subsequent processing which serves to promote memory encoding' (Axmacher et al, 2010).

As a coach or manager you are often on the lookout for new, effective ways to help your clients or colleagues get the most out of themselves. We know that there are strong links between behaviour and the architecture of the brain. For these individuals as designers of their futures this is important for them to be acutely aware of this too.

A fascinating study that is described as the first to correlate social media use with brain activity looked at reward activity in the brain and Facebook (Fareri and Delgado, 2014). The researchers found that they could predict the intensity of a person's Facebook usage by the activity in their NA. Facebook has a feature where people can 'like' information posted on it. This can be thought of as positive social feedback related to one's reputation. Each participant completed the Facebook Intensity Scale disclosing how many friends they had, how long they spent on Facebook and some general thoughts. Next, the participants were interviewed on video and then their brain activity was recorded in various situations. For example, one scenario was that the participant was told that their video interview had been shown to another person and that person thought highly of them. They were also told that another person had received positive feedback. The NA was activated more strongly when the participants received positive feedback about themselves than when they heard that another had received it. This was predictable. What was interesting was that the strength of this difference corresponded to participants' reported intensity of Facebook use. Another situation the participants were put in was to perform a card task to win money. The NA response to the monetary reward did not predict Facebook use.

As is often the way with studies like this, we do not know from the results whether positive social feedback drives people to interact more on social media or if the sustained social media usage changes the way this feedback is processed by the brain. While this study was on social media use... imagine if the brain worked this way for tasks people need to do at work.

The striatum's functions can be broken down geographically. Studies show that as you move from the ventral-medial to dorsal-lateral striatum there is a move from more prominent value encoding to encoding that better reflects associative and sensorimotor functions (Burton, 2015). The ventral striatum is involved in learning to predict future outcomes and informing us of preferred outcomes, while the dorsal striatum is more involved in holding and evaluating information about the outcomes to allow a better action to be chosen more frequently (Kim, 2018). In addition, the striatum has been linked to addiction, memory, promoting active defensive behaviour and the caudate volume has been positively correlated with intelligence (Myslivecek, 2021).

Why is it important to me as a coach/ manager?

The striatum is very important in motivation, reward and salience. It coordinates motivation with movement in both lower- and higher-level functions. For example, it is involved in our ability to inhibit undesirable behaviours in a social interaction. The ventral striatum is involved in rewards. The mesolimbic dopamine system is thought to underline the generation of reward-seeking actions. It is thought to be involved in assessing the value of reward-predictive stimuli (Wenzel, 2015). How we are motivated to do things, the value we attach to things and what we are prepared to do in relation to those things all come into play here. The caudate nucleus is an important part of the brain's learning and memory system. The putamen is also involved in learning.

Neuroimaging studies with functional MRI (fMRI) shows that the ventral striatum responds to reward. The value of the reward is coded for considering the quantity and probability of this reward. The striatum additionally is activated by novel, unexpected or intense things. The common denominator is thought to be salience (the salience of something is the quality by which it stands out relative to its neighbours). It is believed the striatum may also be involved in cognitive processes involving executive function, such as working memory. We can imagine the striatum, in relation to habits or skills, as a sort of learning machine that is dedicated to achieving success in behaviour (Graybiel, 2015). Activity in these neural loops is modified during habit learning and it helps to powerfully control value-based decision making and habits. The NA biases the direction and/or the intensity of behaviours which increases the likelihood that certain actions are committed (Floresco, 2014).

The caudate nucleus is innervated by lots of dopamine neurons, which links with its roles in learning and memory. The dopamine system is known to be involved in feedback processing. When a person is receiving feedback, we can see that the caudate nucleus is activated. The putamen affects reinforcement learning and implicit learning. Reinforcement learning is obviously very important to coaches and managers supporting people to get the most out of themselves and to respond optimally to their environments. Another way the brain learns and adapts involves the mesocorticolimbic circuitry that finds a hub in the NA. It is well positioned to respond to stress signals and transform them into motivated coping behaviours (Castro, 2019).

The NA has also been found to be involved in dealing with uncertainty and competing actions. It is important in situations when there is some sort of ambiguity about either the motivational relevance of an environmental stimuli, or the best course of action for obtaining certain goals. It can sometimes be hard to ascertain the most appropriate way to respond to situations. The NA is activated to help us achieve goals in these scenarios (Floresco, 2014).

One study showed that delayed rewards lose their value for economic decisions and subsequently constitute weaker reinforcers for learning. This study aimed to use short delays to measure fMRI responses in the ventral striatum, a key reward centre of the brain. Other studies based on the dual-process discounting concept show stronger activations for immediate rather than delayed rewards.

Another study looked at the differences that the magnitude of a reward had on people. The experiment involved showing different instruction pictures to monkeys where each image represents a different quantity of fruit juice. The monkeys had to reach to the spatial location where an instruction picture was presented and they would then receive that amount of liquid. The study found that around half of the striatal neurons showed differing response levels depending on the amount of juice they would be receiving. The takeaways from this are that the striatum uses expectations that give precise information about the upcoming level of reward in order to direct general behaviour.

It is worth mentioning that the striatum is activated by rewards in social situations (Baez-Mendoza, 2013). Social information is integrated into social action and reward. This has obvious individual and organizational implications. Activity in the striatum also increases when viewing or thinking about people with whom we have an intimate relationship or who have a high social ranking (Bhanji, 2014). This region lights up in response to both social feedback and monetary rewards, which likely links to socially influenced behaviours. The striatum responds if we hear that popular opinion agrees (rather than disagrees) with our own evaluation of something. Importantly for teamwork, the striatum is more responsive to cooperative behaviour that generates a reward than a greater reward but that is at the expense of another (Bhanji, 2014). We are also highly sensitive to the fairness of others' behaviours. An insight many people would benefit from remembering.

What can I do with a client or colleague now having understood this?

The links here to practical applications are not strong ones as the experimental data is mostly from precise research into the striatum, rather than the bigger-picture perspective that would give direct suggestions of actions. However, it would be fair to suggest that experimenting with using social rewards in a work environment could be useful. Being in the role of a coach in an environment where you manage people or are part of a team gives you a great opportunity. Recognizing people for specific great things they have done has been shown to trigger the reward response in individuals. If a team is working on a big project or task it can be a great idea to break it down into small achievable chunks. Rewarding the achievement of each chunk can help to keep everyone motivated as they progress.

You could dedicate a whole session with a client to looking at the feedback loops that are set up in their life. For example, how do people who work on their own keep motivated? (One way is through their work with a coach who can give them feedback, praise and encouragement.) At work does your client feel they have systems and relationships in place that give them quick feedback? What about in personal relationships, how do they know they love their partner in ways that make them feel loved?

Purely using your coaching skills and knowledge you can help a client to re-evaluate and set up new feedback loops that serve them better. This can lead to a more frequent and useful release of dopamine and reinforcement learning occurring.

You quite possibly already do some things that activate the NA. Whether you work in a corporate environment or elsewhere the power of visualizations is strong. When we visualize something there are real things happening in our brain. Sometimes this frame can help people who think that visualizing is a very fluffy practice. Imagining something pleasing and emotionally powerful to us activates the NA.

A manager in the role of a coach could encourage the person or group they are working with to imagine a scene when a goal has been achieved. What this is not saying is that by imagining something it will magically occur. That is not what we are talking about here. We are focusing on the changes that occur internally within a person when visualizing something that makes their internal environment a more favourable one. People are more likely to be motivated and take goal-oriented behaviours with a boost of dopamine in the NA, and so visualization may make a particular outcome more likely.

In a personal coaching situation, it might be beneficial to explore how often positive memories are being intentionally created between the two people involved. One way to increase frequency could be to plan unusual, different activities. For some people this could be an activity such as taking a bike ride together, going on a cooking course or going camping. The idea is to do something novel and enjoyable together. We tend to remember these kinds of things much more easily than the repetitive mundane stuff. The same principle could be applied to a corporate environment. Team away-days, bonding activities and things that take people out of the familiar office environment can be great experiences. For any coach, varying how you do your coaching sessions could sometimes be to your benefit. Perhaps you could take a walk with your client while discussing something for which movement could be metaphorically useful.

Insular cortex 04

Enables us a degree of awareness and is involved in our ability to be 'in tune' with ourselves.

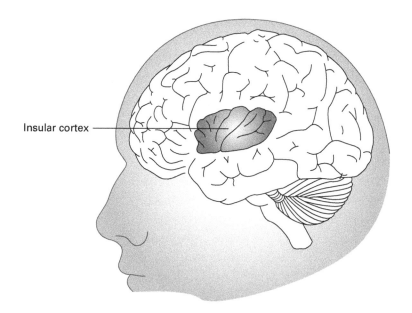

Insular cortex

What is this brain area?

There is an insula in each of the hemispheres of the brain. They are areas of the cortex folded deep down between the temporal, parietal, and frontal lobes. The insular is one of the less well explored regions of the brain with many details of the functional anatomy still elusive (Borzoi, 2016). It is the first region of the cerebral cortex to develop, a process that starts at just six weeks after conception (Gasquoine, 2014).

We do know that it is very well connected to the amygdala (the area important in emotional processes). It is also highly interconnected throughout the brain, which provides the underpinnings for its involvement in a wide range of functions (Borzoi, 2016). The left and right insulae show different patterns of activation and connectivity (Cauda, 2012). The structure of the right anterior insula enables us a degree of awareness, from feeling heat to empathizing with others. Increased grey matter in the right anterior insula correlates with an increased accuracy in the ability to be 'in tune' with your inner body. The insulae are also thought to be involved in consciousness. A meta-analysis of almost 1800 functional neuroimaging experiments suggests there are four functionally distinct regions: 1) A sensorimotor region in the mid-posterior insula; 2) a central-olfactory gustatory region; 3) a socio-emotional region in the anterior-ventral insula; and 4) a cognitive anterior-dorsal region (Udon, 2017).

Why is it important to me as a coach/ manager?

The insulae are involved in a range of things including:

- emotion;
- perception;
- self-awareness;
- decision making;
- cognitive functioning;
- social functioning;
- salience;
- interpersonal experience.

The insular is involved in integrating signals. Autonomic and visceral information comes into the brain and is integrated into emotional, cognitive and motivational functions (Bellucci, 2018). It has a prominent role in the salience network where it assists target brain regions in the generation of appropriate behavioural responses to salient stimuli (Rolls, 2015).

Decision making is another function that the insula plays a major role in (Ibrahim, 2019). Things need incentive values encoded within them. The insula does this by evaluating the stimuli-eliciting subjective feelings states. The rewarding stimulus evoke pleasurable feelings which drive conscious desires for deciding to take specific actions. Aversive stimuli evoke painful feelings which lead to conscious aversions for avoiding particular behaviours. It is in this way that feelings that arise from the insula mediate our behaviours (Namkung, 2017). Decisions when there is a lot of uncertainty are the result of both rational and emotional drives. The insula is likely to play a role in risky decision making (Uddin, 2017).

Antonio Damasio proposes that the insula plays an important role in his 'somatic marker hypothesis' (Damasio, 2008). This concept is much easier to understand than it sounds. It promotes the idea that rational thinking cannot be separated from feelings and emotions. The idea is that we use bodily signals to help us make decisions. Damasio helped put the insulae on the map (as it were) and drive more research into it by suggesting that the insula plays a vital role in processing these sensations so we could use them to help make decisions. James-Lange theory of emotion also relies on the insula. Here the interoceptive information that reflects changes in autonomic nervous system activity levels are received (Gasquoine, 2014). If you were thinking the amygdala was the main player in emotions… you can summarize their responsibilities as follows. The amygdala plays a role in automatic (implicit) responses. It is part of an impulsive system. The anterior insula is involved in subjective (explicit) experiences (i.e. subjective feelings). It is part of a reflective system (Namkung, 2017).

It is proposed that subjective feeling experienced by people involves representation of bodily responses elicited by emotional events. Different people have different levels of sensitivity to internal bodily responses. In one study, people were asked to judge the timing of their heart beating. Through functional magnetic resonance imaging (fMRI) the researchers saw increased activity in the insulae and a couple of other brain regions. The level of activity in the right anterior insula predicted how accurate the individual was in judging their heart rate. Additionally, this study showed that the volume of grey matter in this area correlated with both interoceptive accuracy (our ability to be aware of what is going on inside us – for example, how often our heart is beating) and subjective ratings of visceral awareness. We look at this in more depth in Chapter 27 on mindfulness.

Interoception is a core topic relevant for anyone wanting to raise self-awareness. Interoceptive signals reach the posterior insular first. This is considered the primary interoceptive cortex where low-level sensory features are processed. The information is then passed to the anterior insular, here higher-order interoceptive representations reach our awareness. This is when the right anterior insula steps into its special role in integrating this awareness into conscious emotional feelings. It also plays a role in various motivational, executive, social and self-aware processes. Together with other brain regions they mediate the related functions (Naqvi, 2014).

Linked to self-awareness and the processing of emotions is empathy. Our ability to perceive, understand and experience others' feelings in relation to ourselves. It also requires social cognition. The sensorimotor system for subjective feelings and social interaction. All of which are functions involving the insula (Uddin, 2017).

A Japanese study addressed the effect of looking at images of painful events and images that evoked fear and rest (Ogino et al, 2007). Participants were told to imagine pain in their bodies when they saw the pain images. It was found that, when people imagine pain, areas of the brain involved in pain are activated, including the right anterior insula. It has been proposed that people who are prone to anxiety show an altered interoceptive prediction signal resulting in a prospective aversive body state triggering an increase in anxiety, including worrisome thoughts and avoidance behaviour. The anterior insula plays a key role in this process. Sufferers of irritable bowel syndrome have been shown to have abnormal processing of visceral pain in the insular cortex and this is related to dysfunctional inhibition of pain in the brain.

Another interesting study showed that hypnosis affected the response of the insulae during exercise. Healthy individuals were hypnotized during constant-load exercise to believe they were cycling normally, downhill and uphill. The blood flow to their left insular cortex decreased when they thought they were cycling downhill, and increased in their right insular cortex when going uphill. This and other imaging studies have shown that the insular cortex may play a big role in cardiovascular regulation. It is possible that it might modulate the physical responses based on our perceived levels of effort.

What can I do with a client or colleague now having understood this?

Some of the direct opportunities you have with clients are linked to utilizing the brain's plasticity and improving the functioning of the insular cortex area of the brain. An MRI scanner study showed that people who meditate have significantly thicker right anterior insula; mindfulness meditation has also resulted in positive scanning results showing an improvement in cortical structure. Mindfulness meditation involves a specific non-judgemental awareness of present-moment stimuli without cognitive elaboration. The fact that the brain's structure can change is very encouraging to coaches seeking to help people change and develop. For clients who are wishing to gain the benefits derived from an increased interoceptive awareness of body states then working on the connectivity of their insular cortex could be a good move. This can be done through various forms of mindfulness training. Mindfulness is covered in more depth in Part Six.

Some research has pointed towards yoga and pilates as also being good for increasing insula activation. Perhaps in a corporate environment a class could be arranged on site after work once a week. Internal coaches and managers in coaching roles could lead by example and attend, or even start meetings with five minutes of yoga! In personal coaching, perhaps you could invite clients to experiment for one or two months to see for themselves whether they experience enough knock-on benefits to merit the time investment.

The benefits of increased self-awareness are well documented in fields such as organizational leadership, management and emotional intelligence. If coaching is happening as part of a 360-degree review, then the opportunity to explore this is obvious. What would be different if you had such a level of self-awareness that you knew clearly how others perceived you?

The big opportunity when working with any human is to help them understand they have more influence over their subjective experiences than was once thought. They can condition themselves to have more helpful emotional responses to situations. It is possible to become more self-aware, more empathetic, and more emotionally intelligent.

Amygdala

Known for being part of the limbic system that is heavily involved in regulating our emotional life.

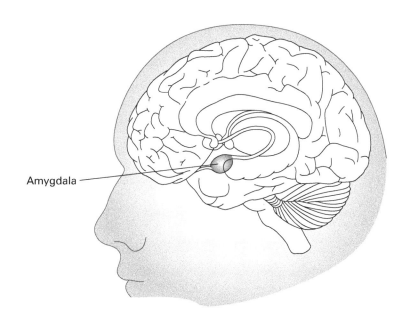

Amygdala

What is this brain area?

There are two amygdalae and these almond-shaped groups of sub-nuclei can be found within the anterior medial temporal lobes. They are known as part of the limbic system (the controversial group of brain areas that are heavily involved in our emotional life – some neuroscientists suggest that we abandon the concept of a functionally unified limbic system, as the underpinnings are no longer accepted as accurate). We do know this region is evolutionarily old, assumed to contribute to self-preservation (Andrewes, 2019) and that its functions and circuitry have been well conserved over the years (Diano, 2017).

The amygdala is highly interconnected with other regions within the brain. For example, the vmPFC, AI, dmACC and amygdala work together as the salience network. The amygdala is also associated with various neurotransmitter systems. For example, dopamine plays critical roles in both modulating neurons and affecting memory formation. Concentrations increase in the amygdala during stressful conditions and the activation of dopamine receptors in the amygdala are necessary for encoding fear-related memories (Lee, 2016). Predictably, social functions in the mPFC-amygdala pathways may be under oxytocinergic influence (Gangopadhyay, 2020). The amygdala is also prominently modulated by serotonin. Genetic and pharmacological alterations of the serotonin transporter appear to affect amygdala activation in response to emotional stimuli (Bocchio, 2017).

What we see in the amygdala is also what every manager dreams of. A team of flexible individuals who can navigate the complex, ever-changing environment. At the neuronal level we know that individual neurons do hold membership in multiple, functionally distinct, neural ensembles. Their abstract or high-dimensional representation enable this flexibility around both social and non-social behaviours (Gothard, 2020).

The amygdala responds to environmental stimuli that may be picked up consciously or unconsciously. It is specifically involved in motivationally relevant stimuli such as fear and reward. When we are anxious our amygdala gets activated. However, we must remember that the amygdala detects all emotions. It processes them in order of significance, and fear is very significant. It is connected to the cognitive parts of the brain, specifically the prefrontal cortex and anterior cingulate cortex (ACC). So strong are the links that when the amygdala becomes unsettled for any reason often so does the ACC. When the amygdala is activated through anxiety, the knock-on effect is that areas of the prefrontal cortex and ACC become disrupted, meaning that short-term memory, attention, and our ability to make risk benefit assessments are all affected.

A further critical example of the highly interconnected nature of the amygdala relates to fear and threat. (Remember that although this is important, it isn't by any means the only role of the amygdala!) You may have heard about the amygdala in relation to the adaptive fear response. This essential behavioural adaptation that keeps us safe is fairly well written about. If we see a file on the desk of our boss with our name on it, or what we thought was our name on it, this sensory information is quickly passed to our thalamus. The thalamus doesn't know for sure whether this information means we are in danger or not, but forwards the information to the amygdala. The amygdala takes action to protect us and tells the hypothalamus to initiate the fight-or-flight response. We realize that our heart rate and breathing rate have quickened; the slower, but equally important process that is also initiated when we see the file involves the thalamus sending the sensory information to the sensory cortex, where its meaning is interpreted. The sensory cortex identifies that there are multiple interpretations of the data and sends it to the hippocampus. Here, previously encountered stimuli and scenarios are taken into account. We consider whether we really saw our name, whether it could have been a harmless file, or even that it was a file we gave to our boss ourselves. The hippocampus concludes that there is no danger and sends a message to the amygdalae to tell the hypothalamus to shut off the fight-or-flight response. The amygdala is certainly involved in influencing the constellation of sympathetic autonomic preparations for the fight, flight or freeze response. This extends to detecting fearful and other negative facial expressions, including the more subtle ones like untrustworthiness and jealousy (Andrewes, 2019).

Sometimes a fear response is innate, other times it has been conditioned. Whenever we are, in effect, learning to fear something, the amygdalae are thought to be involved. The study that led to the identification of the neuronal circuits used mice. These mice first went through a simple behavioural task that conditioned the fear response to a sound. Once this was set up, and a fear response was observed, the researchers used advanced pharmacological and optogenetic techniques (enabling the monitoring of individual neurons). They showed that both the medial and central nuclei of the central amygdala were involved in either learning or behavioural manifestation of the fear responses. By manipulating different parts of the amygdala they could identify the subdivisions that seem to be responsible for learning to be afraid and acting afraid.

This study was even able to identify the neurons within the structures that were responsible for the components of the fear response. The relevance of this is huge. People who suffer from debilitating fear responses, such as many of the anxiety disorders or post-traumatic stress disorder, could all benefit. Further research is needed into how the selective manipulation of these neuronal circuits can be achieved most efficiently. Any coaching tools that claim to help people overcome an undesirable fear response would likely need to work on these circuits.

It is exciting to see coaches realize just how interconnected our experiences are with shaping our brains. Anxiety is a common and, to a degree, healthy response to stress. In short bursts it helps us process and cope. When anxiety continues over a prolonged period of time, however, problems can arise. The link between what is happening in our lives and what is happening in our brains is fascinating. Studies that have looked at adults with anxiety disorders have shown that they have enlarged and highly connected amygdalae. Their brains have changed in response to their experiences. Studies with lab animals that were put in environments that caused chronic stress have showed that these animals' amygdalae grew additional synapses. They also increased their synaptic connectivity.

Another study focused on children who suffered from anxiety (the specific levels were lower than what is considered clinical anxiety, however). They found that four functional neocortical systems in the basolateral amygdala were affected. The systems that deal with perception, attention, and vigilance, reward and motivation, and detection of salient emotional stimuli and regulation of emotional responses were all affected. This was seen in children aged seven to nine. The changes in these children's brains could predispose them to anxiety disorders later in life. Thus, it is important to keep our focus – as coaches and managers – on the fact that the brain can change.

The amygdala has been shown to be important in a rich and varied social life within humans. Previous research has shown that primates who live in larger social groups have larger amygdalae. In 2010 research also revealed that the volume of the amygdala positively correlated with both the size and complexity of social networks in adult humans. This holds true for both men and women, both younger and older. No such correlations were found between other structures in the brain. The suggestion is that evolutionarily the amygdalae could have evolved, at least partially, to deal with our increasingly complex social life. This 'social brain hypothesis' is a popular one, although less well understood. We might come to understand the amygdala as helping to fine-tune and add nuance to social behaviours. It has been seen playing a role in:

- social decision making;
- observational learning and vicarious reward processing;
- face, gaze and eye contact processing;
- representing social status;
- allocating special attention to social stimuli;
- social anxiety;
- producing facial expressions;
- social memory;
- coordinating autonomic responses elicited by social stimuli (Gothard, 2020).

Why is it important to me as a coach/ manager?

The amygdala is important in:

- emotional responses and these drive a lot of our behaviour;
- memory;
- attention – the ability to focus in on something while excluding others;
- social processing.

Each of these things are of regular importance to working with people, as you may already recognize, and we'll see more as we progress through this book. Coaches and managers who understand that actual changes take place in the brain as a consequence of experience are in a better position than those who do not. This fundamental point has the power to change how you approach your work with people. Expecting change to occur quickly, even once a stimulus has been removed, is unwise. The brain needs time to rewire again.

In a corporate environment it can frequently be the case that people get scared about something. They aren't likely to articulate that they are having a fear response, but they happen more regularly than we'd like to think. It doesn't have to be a full-on scary situation to have an internal effect. This could compromise the High-Performing Neural Environment quite easily. Perhaps a reshuffle is announced, and people are scared about whether their role will be changed or removed altogether. Perhaps end-of-year figures come in and someone knows their department won't have performed as well as last year. Perhaps it is as simple as being requested to attend a meeting with the boss. As a manager or leader in a coaching role it is useful to remember that it will be hard for an individual to process much else until their fears are under control. It is a very powerful emotion and can overpower other emotions and processes.

Optimism can replace fear. This is great news and often can be used to our advantage. Diverting a focus from what may happen that is negative, such as some form of telling off from a boss, to what is realistically optimistic – such as an opportunity to learn, share insights and move forward – can be useful. It is worth consulting Chapter 26 on optimism to ensure that any actions along these lines are in the best overall interests of the people we work with.

Enhancing trust can also decrease amygdala activation. Creating opportunities as a coach or manager to increase perceived trustworthiness and trusting your client is always a valuable investment. Coaching those you work with to do the same with others is also a great plan. They may not be aware that to get the best out of someone whose amygdalae are highly activated one of their options is to build trust with that person.

What can I do with a client or colleague now having understood this?

- Be alert to a client letting fear hold them back. Question whether their amygdalae could be overreacting. Often even by posing this question you can open a dialogue with a lower fear response in place.

- Just by being around fear-inducing stuff your unconscious can pick it up, through emotional contagion. Therefore, consider experimenting with reducing these inputs, egturn off the news, don't read a newspaper, avoid negative people.

- Deal with things – of course, this sounds obvious to a coach! Helping your client to look at challenges and things that scare them is hugely important because it frees up other areas of their brain to work effectively and efficiently.

- Help your client to explore unconscious fear triggers.
- Don't allow your people to make big decisions when they are anxious, because they are unlikely to be considering the whole picture.
- Despite what people say, working under negative pressure doesn't create optimal brain environments. Encourage experimenting with alternatives.
- Encourage people with a negative attributional bias to reflect on the positives.
- Encourage a healthy gut microbiome... the research is stacking up that a healthy gut may contribute to a range of amygdala-dependent brain functions, from pain sensitivity to social behaviour, emotional regulation and therefore mental health (Cowan, 2017).

Anterior cingulate cortex

Primarily responsible for conflict or error detection.

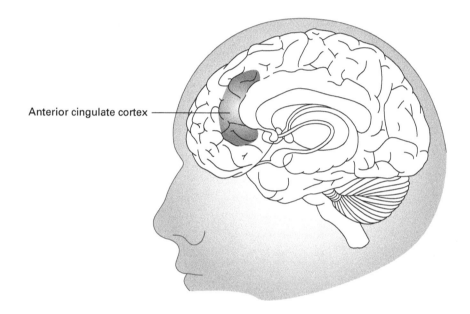

Anterior cingulate cortex

What is this brain area?

The anterior cingulate cortex (ACC) can be found looking like a collar around the frontal part of the corpus callosum (the bundle of neural fibres that connect the left and right hemispheres).

We now know that the ACC is involved in many different functions. This list includes:

- Motivation, decision making, learning, cost-benefit calculation, conflict and error monitoring (Apps, 2016).
- Social behaviour including judgement and decision making, social affiliation, motivating behaviours that promote cooperation, trust and altruism (Touroutoglou, 2019).
- Pain processing, performance monitoring, value encoding, emotion and learning (Shenhav, 2013).

Having said all that, the dACC still remains quite a mystery (Heilbronner, 2016).

Cognitive Control

The cingulate cortex overall may serve as a functional hub of communication that synchronizes and integrates information from otherwise segregated systems to drive goal-directed behaviours (Touroutoglou, 2019). There is strong evidence that the dACC is involved in control and persistence (Heilbronner, 2016).

A major early theory of ACC function is in error detection and self-correction. This region can integrate both errors and rewards over time to develop a representation of the value of competing options. Evaluating cost-benefit information influences how motivated we feel and what our decisions are in terms of maintaining or switching behaviours (Apps, 2016). It is involved in dealing with conflicts between autonomous and cognitive processes.

Error detection

The classic activity that supports the idea of error detection involving the ACC is called the 'Stroop task'. This activity involves a person naming the colour of a word. The tricky bit is that the words are of colours, and sometimes the colour matches the word (red appearing in red ink) but sometimes they don't match (red appearing in blue ink). The test is considered to measure selective attention and is used in both psychological and neuropsychological circles. Both the ACC and the dorsolateral prefrontal cortex (dlPFC) are activated during this task. The ACC is responsible for selecting the correct response and focusing attention appropriately.

The ACC also has roles in many autonomic functions that are part of the control system that acts mostly below the level of consciousness to control things such as blood pressure and heart rate regulation, in addition to rational cognitive functions such as decision making, empathy, impulse control, emotion and reward anticipation. The dorsal and ventral parts of the ACC are connected to different other areas of the brain and their functions reflect these connections. The dorsal part (dACC) is connected with the prefrontal cortex, parietal cortex and motor system. The ventral part of the ACC (vACC) is connected to the amygdala, nucleus accumbens, hypothalamus and anterior insula. The vACC is involved in evaluating the salience of emotion and motivational information.

The ACC has a lot of a particular type of neuron called 'spindle cells'. These cells have only one dendrite. We believe them to be important in the intelligence levels of humans as they are linked to the quick processing of information in other mammals. The ACC may be involved in linking between the conscious and unconscious.

Mental fatigue

Several studies have highlighted the involvement of the ACC in mental fatigue (Choo, 2005). Almost everyone experiences mental fatigue or exhaustion at one time or another. At the National University of Singapore, researchers used functional magnetic resonance imaging (fMRI) to look at the brains of men and women aged 19 to 25. These individuals were deprived of sleep for 25 hours and given a simple task repeatedly throughout that time period. It was found that this gradual increase in mental fatigue correlated with decreased activity in the ACC. We know this area to be important as an interface between motivation, cognition and action. It is also linked to the utilization of reinforcement information to control behaviour. So during mental fatigue when we feel lethargic and our thoughts feel slow it is possibly because our ACC is suffering from decreased activation!

Another study looked at the actual way that meditation can reduce anxiety. Mindfulness meditation was the style used here and the participants of the study all had no previous experience. They went through four 20-minute classes to learn some mindfulness meditation. As you'll read in Chapter 27, this involves focusing on breathing and sensations within the body. It also means non-judgementally evaluating any distracting thoughts or emotions. The scanner that looked at the participants' brains before and after they meditated did arterial spin-labelling MRI. The participants were also asked about their anxiety levels before and afterwards. The best results showed a reduction of 39 per cent in anxiety levels. This correlated with increased activity in the ACC and the ventromedial prefrontal cortex (this area is involved in controlling worrying and executive-level function).

Social behaviour and observational learning

The ACC plays a crucial role in evaluating the behaviours of others and estimating their level of motivation (Apps, 2016). Obviously this is a great skill for managers. The neural specialization you find in the ACC for processing social information along with how well connected it is to other structures with known social-cognitive functions makes it important in observational learning (Touroutoglou, 2019). The dACC and the AI have been repeatedly noted to be involved in affective empathy. They both respond to both an individual's pain, and the perception of it in others (Molenberghs, 2018). The ACC have even been identified as important in facilitating a person's ability to recognize and understand his or her own emotions often referred to as emotional awareness (Smith, 2019).

Why is it important to me as a coach/ manager?

ACC is often referred to as the conflict monitor or detector. As you've already seen, the ACC connects to different parts of the rest of the brain and is multifunctional.

Being able to talk to a client or colleague in terms of the effect that mental fatigue is having on their brains can sometimes add a degree of distance to problems, which can make them easier to handle. Many of the coaches and managers we've trained over the years are grateful to be able to explain things rationally and logically. As a manager, if you notice that a team member is struggling with mental fatigue regularly, and it might affect their health and well-being, then a conversation could be useful. How this is approached is obviously of great importance. Simply starting by sharing the above study could be a good way to illustrate that it isn't necessarily any-thing a person is doing wrong or should be pushing through. Rather, it is more like taking care of a machine that has certain limitations.

Something else of interest is a variation on the original Stroop task in-volving a mix of negative emotional words such as 'grief', 'violence' and 'pain' along with neutral words such as 'clock', 'door' and 'shoe'. The words are in various colours and the aim is to name the colour of the word. Many experiments have shown that people who are depressed take longer to name the colour of the negative words than the neutral ones. This highlights the conflict between the emotional relevance to the individual (dependent on their mental state) and the word.

Often, we learn a lot in biology about when things don't work normally. Another area of research to be aware of when working with people is obses-sive compulsive disorder (OCD). The number of people suffering from vari-ous mental health challenges has risen. OCD tells us more about the role of the ACC. In people with OCD low levels of glutamate activity have been noticed in the ACC whereas, in other areas of the brain, high levels have been present. Lower grey-matter volumes have also been noted in the ACC. In other pathologies when lesions are present in the ACC people have diffi-culty performing the Stroop task, exhibit emotional instability, have an inability to detect errors and experience inattention.

What can I do with a client or colleague now having understood this?

The ACC may turn out to be the centre of free will in humans. Francis Crick (co-discoverer of DNA) proposes the ACC as a front-runner candidate. With this area of the brain there may be less to 'do' on an interaction-by-interaction basis and more to have an overall awareness of.

There is a great experiment that involves a screen, a dot, and a black box in the centre of the screen. The dot travels from the left to the right of the screen at a constant velocity. When it reaches the black box, it disappears behind it. When it emerges, you have to state whether it travelled behind the box at the speed you expected or more quickly or more slowly. The ACC is instrumental in detecting this potential unexpected occurrence.

Expectations are powerful things, and our ACC is expert in matching whether what we expect is what we get. If we speak to our boss after just handing in a report we've worked really hard on, and are expecting a 'thank you' or some praise for efforts but instead get a brush-off, then we need to adjust.

Our ACC is dedicated to drawing our attention to what is different from what we expect. Perhaps you've had the experience of receiving an e-mail from a colleague and something feeling not quite right. Maybe you brush it off and move quickly on to the next thing on your to-do list. It may be that your ACC had picked up on something important. Perhaps you walk into a meeting and something just isn't what you were expecting.

One of the gifts from the ACC is to pay attention to error detection. We also know that the ACC plays a general role in behavioural adjustments. Things like credibility strongly influence the judgement of a person about whether something is true or false. Economic decisions can also depend strongly on framing. Some studies provide evidence that the framing induced judgement biases are linked to emotional information that is integrated into any cognitive decision-making processes.

Another ACC-friendly option is to reassess a situation. How we interpret a situation – as coaches we know this – is in a large percentage down to us. Utilizing your skills as a coach to reappraise or reassess a situation can be very useful. Managers need to be experts in managing expectations.

Hypothalamus 07

Responsible for various metabolic processes and synthesizing and secreting neurohormones.

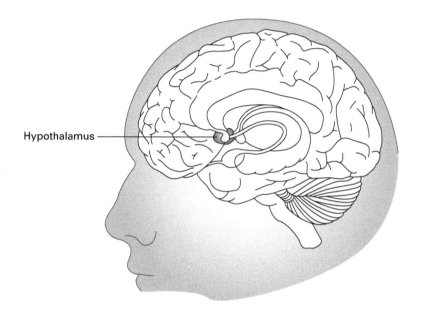

Hypothalamus

What is this brain area?

The hypothalamus is a small area of the brain whose roles include linking the nervous system to the endocrine system via the pituitary gland. The hypothalamus is responsible for various metabolic processes and synthesizing and secreting neurohormones. It is one of the brain areas that it is important to know about for a complete picture, but rarely would direct reference be made to it during a coaching session.

The part of the hypothalamus called the ventromedial nucleus is responsible for the control of food intake. There are several hypotheses that could explain the regulation process. One of these is that adipose tissue (fat, to most of us) produces a proportional signal that acts on the hypothalamus to decrease food intake and increase energy output. Evidence supports leptin (a hormone) acting on the hypothalamus in this way. Another hypothesis is that certain gastrointestinal hormones inhibit food intake. These hormones are released when food gets into the gastrointestinal tract (the stomach being a part of this) and signal to the brain to produce feelings of satiety. A third hypothesis suggests that glucose utilization in the neurons affects our natural food intake. So when we have low blood glucose levels we feel hungry. Another theory suggests that a decrease in body temperature below a set point makes us feel hungry, and above a set point inhibits appetite. Ever felt hungrier in winter and less interested in food in summer?

Why is it important to me as a coach/ manager?

The hypothalamus is important in regulating hunger, aspects of parenting and attachment behaviours, thirst, fatigue and sleep. It is considered part of the limbic system and connected to the amygdala. It responds to olfactory stimuli (including pheromones) and stress. It is also involved in the mediation of emotional responses.

For any coaches working with people trying to lose weight, due attention should be paid to how our brains create and respond to hunger and satiety. Some of the hypothesis will fit with your foundational understanding of hunger; however, in recent times with a variety of diets hitting the marketplace sometimes these basic principles seem to get a little lost. If a client is struggling, it may be worth returning to basics and looking at what is happening. It would also be wise to bear in mind what we know about willpower and self-control. Look at Chapter 25 on willpower for some ideas.

What can I do with a client now having understood this?

There are not any obvious things you can do with a client having understood the primary functions of the hypothalamus unless you are perhaps working on weight-loss goals.

Hippocampus 08

Has roles in the consolidation of information from short- to long-term memory.

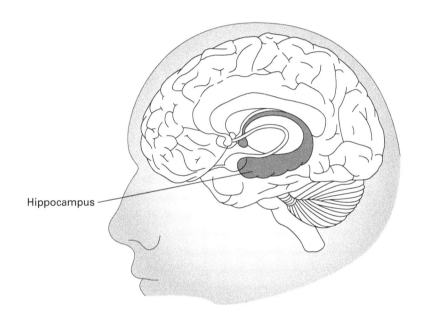

Hippocampus

What is this brain area?

The hippocampus can be found in the medial temporal lobe of the brain. Visually it resembles a seahorse and its name, in Greek, comes from components meaning 'sea monster'. It is special because it is one of the most plastic regions of the brain. Neurogenesis (the creation of new neurons) occurs here (Jurkowski, 2020).

It is considered part of the limbic system (or midbrain) and has roles in the consolidation of information from short- to long-term memory. Long-term potentiation (LTP), one of the mechanisms by which memory is stored in the brain, was first discovered in the hippocampus and has been studied frequently in that area of the brain (a great area to study LTP because there are clear layers of densely packed neurons). LTP involves altering the strength of connections between neurons (we'll learn more about this in Chapter 17 on neuroplasticity). Donald Hebb is credited for the roots of the simplified phrase: 'Cells that fire together wire together' (Hebb, 2002).

The hippocampus is essential for learning. It is involved in the acquisition of new memories and the retention of acquired memories (Avchalumov, 2021). The most well-known function of the hippocampus is episodic memory. This is the concept of first-person experienced events within the context of space and time. For example, what happened in the meeting yesterday, who else was there and where did I sit in relation to them (Buzsaki, 2018). As has historically been the case in medicine, we've learnt a lot about how the hippocampus worked by studying those who had lesions or disease in the brain region. Patient H.M. had large portions of the medial temporal lobe, including the hippocampus removed. He suffered server anterograde amnesia for facts and events (declarative memory). But his implicit long-term memory was not affected, and he could continue to learn procedural skills (Weibacher, 2016).

A (now famous) study looked at taxi drivers. This study was carried out at University College London by Eleanor Maguire and colleagues in 2000 (Maguire et al, 2006). It was known that people's hippocampi are more active when they are successfully navigating, as this had been tested in simulation tasks. It is also thought to be involved when we find shortcuts or alternate routes. The study at UCL looked at black cab drivers' brains and compared them to normal drivers' brains. What they saw was that the cab drivers had bigger hippocampi. The question then becomes, are people with bigger hippocampi naturally drawn to jobs where they are naturally talented – or does navigating and finding shortcuts daily make the hippocampus grow bigger? This study did show that the amount of grey matter correlated with the length of time the cab drivers had been in this role. (Specifically, what was found was that the posterior portion of the hippocampus grew at the expense of the anterior portion.) Further studies explored this question and others and found that it was the type of attention and practice that cab drivers are required to do that changes their brains. Not that they are born with navigation superior brains.

Interestingly, rather than a correlation between the overall size of the hippocampus and an individual's ability to recall things, a recent study showed that it was instead the relative contributions of the posterior and anterior parts of the hippocampus. In tests with normal individuals, it was found that a larger posterior hippocampus, and a smaller anterior hippocampus, was linked to better recall. It has also been shown that the link between the posterior hippocampus and recollection depended on interactions with other parts of the brain. The ventral hippocampus is involved in regulating emotional affect and memory (Bryant, 2020).

The hippocampus is involved in emotional processing. The ventral part has bidirectional glutamatergic communication with the amygdala (Bryant, 2020). The projections to the PFC are thought to act as a fear gating mechanism. The updated thinking is that the ventral hippocampus is a context dependent decider of behavioural strategy. So not as black and white as just an inhibitor. It is also involved in mood regulation through its connections to the amygdala, nucleus accumbens and hypothalamus (Jurkowski, 2020).

There is some evidence that post-traumatic stress disorder (PTSD) and cortisol levels can affect the size of the hippocampus. For example, a study looked at 15 children aged 7–13 who were suffering from PTSD over a time frame of 12–18 months. They measured the children's bedtime cortisol levels (which is a marker of stress) and found that those with more severe symptoms and higher levels of stress also had reductions in their hippocampal volumes. We know that cortisol, from the hormonal group of glucocorticoids, has been shown to kill hippocampal cells in animals.

A study on elderly people over a 10-year period gives us confirmation that the hippocampus decreases in size because of depression (Heijer et al, 2011). The study concluded after scanning individuals that people with a smaller hippocampus to start with were not at higher risk of developing depression. Instead, it is the depression that causes a decrease in the size of the hippocampus.

Noradrenaline, from the locus coeruleus, is involved in modulating the encoding, consolidation, retrieval and reversal of hippocampus-based memory.

As is dopamine. The two neuromodulators work together to influence long-term synaptic plasticity and memory (Hansen, 2017).

While the hippocampus is undoubtably involved in memory, recent research suggests it is more accurate to think of it as performing multiple computations on a specific representation (Lisman, 2017). Rather than just being involved in encoding and retrieving episodes from memory it is now believed it plays a more general role where it builds flexible representations that span multiple experiences, are sensitive to goal states and guide novel decisions (Mack, 2018). There are at least four processes it is central to:

1 Enabling bridging across shifts in context, so allowing a person to recognize when they have returned to a familiar context and where they are within that context.

2 Providing a map that associates internal and external cues. This is useful for...

3 searching processes on the map. For planning, deliberation and identifying novel paths and connections.

4 Providing a resource for consolidation. Essentially it helps us construct cognitive maps. You can think of it as an integrative code building that binds together elements that share spatial, temporal, or conceptual features to form codes that relate to each other and capture commonalities of our experiences.

A great question is often: 'Can we improve this area of the brain?' A linked question when we talk about areas that are involved in memory is often: 'Does it have to go downhill as we get older?' Research helps us answer both of these questions. The short answer is that 'yes' we can improve hippocampal function and 'no' your memory doesn't have to go downhill as you age. Studies have reliably shown a relationship between the size of the hippocampus and memory performance. A study showed that aerobic exercise could increase the size of the hippocampus in adults (this study involved people 55–80 years old) (Erickson et al, 2010). Another study looked at physical fitness in children aged 9–10, which showed that those who were fit performed better on memory tests and had larger hippocampal volumes.

As is often the great thing within science, sometimes studies show one thing and then another study shows something to the contrary. It often takes time and many studies to get to a conclusive point. This is just something that we need to be aware of and exercise caution around. So, the proviso to the above statements are that some studies have shown a loss of neurons in the hippocampus of elderly people, but further studies, using more precise techniques, showed minimal differences between older and younger people. Again, some MRI studies have shown the hippocampus decreasing in size, while others have not shown this.

Why is it important to me as a coach/ manager?

The hippocampus is really involved in how we perceive what goes on. Take its role in scene construction theory for example. Here it acts to facilitate the construction of atemporal scenes. This allows the event details of episodic memories and imagined future experiences a foundation where they can reside. This is linked to people's ability to predict the future (and so evaluate whether to do something) and even dreaming and mind-wandering which are critical for innovation (Ekstrom, 2018).

What people remember and what they forget has potential ramifications on their levels of fulfilment and ability to reach their goals. It also affects their memory guided decision making. It has even been suggested that fundamentally the purpose of memory is to guide future decisions! (Biderman, 2020).

The hippocampus has high levels of glucocorticoid receptors (cortisol falls into this category of hormones) and so it is susceptible to long-term stress. Exposures to stress or elevated cortisol levels impair performance on memory tasks that depend on the hippocampus (Kim, 2015).

In an organizational situation people can often be resistant to change. We often find that people's memories of what has gone before in that organization are very important to them. Being aware of a few things around memories can be useful. First, just because something happened does not mean everyone will remember it the same way. This seems obvious, but is often forgotten when people are defending their recollection of something. Second, people seem to develop an attachment to what has gone before and can seem to take it as a personal criticism if there is a suggestion of how things can be improved in the future. It might be as simple as giving people the opportunity to voice their historical experience and perspective on the new suggestions.

What can I do with a client or colleague now having understood this?

There are several things that a range of studies tell us are wise moves:

- sleep;
- regular aerobic exercise;
- stress reduction to keep cortisol levels from getting chronically high;
- give your global positioning system (GPS) a break and engage your brain.

While the hippocampus is vulnerable to the aging process, there are things that can be done to counteract the age-induced alterations in hippocampal signalling, structure and function. These include:

- caloric restriction;
- physical exercise;
- environmental enrichment.

(Bettio, 2017)

Cerebellum

Coordinates movements; controls posture, balance and fine motor movement; involved in motor learning.

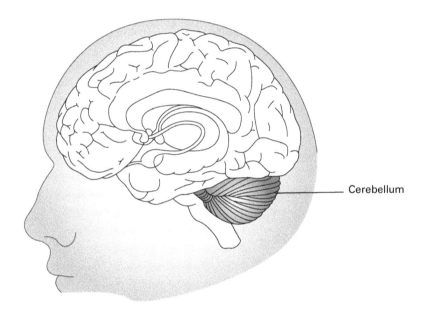

Cerebellum

What is this brain area?

The cerebellum is also known as little brain due to its size and location (Amore, 2021). You can find it as part of the hindbrain and it weighs about 11 per cent of the whole brain weight in adults and uses approximately a fifth of the energy budget (Diedrichsen, 2019). However, it contains almost 80 per cent of the global brain neurons and is involved in far more roles than we used to think. It is now known to be important in sensorimotor, cognitive, and affective functions. The cerebellum is one of the phylogenetically oldest brain structures in all mammals and one of the earliest structures to develop in the central nervous system (Ceylan, 2015).

The cerebellum is involved in:

- motor control (Baresm, 2018);
- predictions (motor and non-motor) (Sokolov, 2017);
- coordination (Peterburs, 2016);
- performance monitoring (Peterburs, 2016);
- cognition – including spatial reasoning, language, working memory (Wagner, 2020);
- most tasks! (attentional control, emotional processing, social cognition) (Diedrichsen, 2019);
- mood regulation (Leggio, 2018);
- processing positive and negative stimuli (Leggio, 2018);
- increased impulsivity and aggression (if damaged) (Klaus, 2021);
- mentalizing (Overwalle, 2015).

Why is it important to me as a coach/ manager?

The cerebellum interacts closely with the rest of the brain. Traditionally it has been studied in isolation, but the more recent research is showing it to be a structure that adjusts processing in much of the brain (Bares, 2018). This is a trend we have seen across so much of neuroscientific research. Whilst in the early days with the new technology we started to uncover the functions of brain regions in isolation, now we see how interconnected the reality of functions are. For any coaches working with children or adolescents, or managers supporting parents, this knowledge is priceless. Until the brain is fully structurally developed (around age 25) it is like you are dealing with an internal building site. There is mess, disruption, new layouts trying to be established. This affects the services that can be expected. One of the big mistakes we see people making in dealing with young people is expecting them to be able to function the same as adults, without having the hardware or software fully installed yet. In technical terms... functional connectivity MRI studies have shown that, via cortico-ponto-cerebellar and cerebello-thalamo-cortical loops, the majority of cerebellum projects to many cerebral association and limbic areas, including the prefrontal and parietal cortex, amygdala, hippocampus, hypothalamus, striatum, and brain stem (Lange, 2015).

The cerebellum and cerebral cortex increase in mass together during development. This is suggestive of a concerted evolution of the corticocerebellar circuits and of the behavioural/cognitive (and not only sensorimotor as previously thought) functions they mediate (Leggio, 2018). It is now believed that the neocortex-cerebellum circuit acts as a joint dynamical system both in classical sensorimotor contexts and reward-related, cognitive processing (Wagner, 2020). So when things go wrong you see problems outside of motor processing also. Executive functioning, spatial cognition, language and even changes in personality and affect (Lange, 2015).

Some psychiatric conditions are linked to damage in the cerebellum. Autism spectrum disorders, deficits of language processing and vocal learning, schizophrenia, bi-polar disorder, major depressive disorder, anxiety disorders, and temporal processing impairments (Hull, 2020 and Phillips, 2015). Many people now are diagnosed with one of these conditions.

Emotional regulation is a topic that has wide ranging implications. It is a skill set that most people require in various aspects of their life. So, knowing when this is reduced in people is useful. The anatomical evidence shows that the connections between the cerebellum and areas of the brain involved in emotional regulation are bidirectional. This means the process can be influenced and adapted in multiple ways.

What can I do with a client or colleague now having understood this?

A very helpful piece of research from the Human Connectome Project in 2015 found that the more positive a person was affected how well their brain was connected. A well-connected brain brings all sorts of functional benefits... so if you can help you clients or colleagues improve this, they are likely to find many other improvements start materializing.

Specific, evidence-based things you can try include:

- physical exercise (increases blood flow all over);
- meditating (overall brain connectivity);
- reading complex things (increases neuronal connections and speeds up mental processing);
- learn a musical instrument (cognitive and physical coordination very good neural workout);
- learn a foreign language (stimulates brain and expands memory);
- travel (use spatial navigation, new stimuli);
- take up a new hobby (ideally one that requires some physical coordination and new thought processes).

PART TWO
Brain chemicals

Discover the communication molecules of your mind

We have explored the key areas of your brain that you need to be aware of, and now we turn our attention to the chemicals that are at work in this amazing organ. What actually are brain chemicals? We use the term here to encompass a range of chemicals that have an effect on the brain. Some also have effects in other areas of the body. To get really sciency for a moment, let's zone in on neurotransmitters, which are the chemicals that *transmit* a signal from a neuron (a brain cell) to another cell across a synapse (a gap). As we look at more of these basic terms in Part Two and Part Three the picture will get clearer and your understanding will increase. A great way to learn anything new is to layer-in information. During our programmes often participants find it reassuring that knowledge of the whole picture is not expected to be crystal clear at the start.

So we can imagine these neurotransmitters to be packaged up into little balls called synaptic vesicles. When instructed, they are released into the synaptic gap so that they can bind to specific receptors on the other side of the gap. If you did A level or A2 biology you would have probably covered the lock-and-key mechanism that we think of as underpinning receptors binding with neurotransmitters. Neurotransmitters include glutamate, dopamine, serotonin, oxytocin, adrenaline and noradrenaline.

Neurotransmitters such as dopamine, serotonin and oxytocin can also act as neuromodulators. This means that they are secreted by a small group of neurons and move through the nervous system affecting lots of neurons.

Some chemicals we cover in Part Two are not neurotransmitters, they are hormones. Cortisol is one such example. It is included here because it has such important actions within the body and on processes linked to the brain – and for a good overview it needed to be mentioned. In fact, we'll take a look at cortisol first.

Caution

There have been a lot of articles and people now talking about different chemicals. Some people get confused.

One of the biggest complaints I hear from sceptics of the concept of using neuroscience to help people is that it gets misused. This appears to be most people's biggest challenge. This caution is included for you because there are people who will just tag on '… *because of dopamine*' to anything in the hope of sounding more credible. I understand the temptation but I don't think it will help in the long run.

Cortisol

Famously known as 'the stress hormone' but also vital for 'housekeeping' for healthy functioning.

What is this brain chemical?

Cortisol is a steroid hormone that is secreted by the adrenal glands (you'll find these just above your kidneys). Generally known as 'the stress hormone' because it is released in response to stress, it has a wide range of functions in the body, including:

- glucose metabolism;
- blood pressure regulation;
- blood sugar level maintenance (through insulin release);
- immune function;
- cognitive function (Adam, 2017, Clow, 2020).

Why is it important to me as a coach/ manager?

In simplified terms, cortisol is involved in both eustress (good stress) and distress (bad stress). Eustress can be thought of as the motivating type of stress that can help you feel challenged, productive and motivated. Distress feels like there is tension and tends not to feel positive. Short-term elevated cortisol levels can affect efficiency, effectiveness, productivity, interpersonal skills, mental state and much more. This corresponds with eustress. It is acute in nature and considered adaptive. Prolonged elevated cortisol levels can impact quality of life (lowered immunity so potentially getting sick more often, higher blood pressure and increased abdominal fat) and impair cognitive performance. This is the distress we tend to be more aware of. Eustress is linked to a heightened state of arousal that primes you to act.

Once this action has been taken your cortisol levels can return to normal. Distress is not so linked to action and often this results in the cortisol remaining in the body – and us feeling generally anxious.

Typically, in response to a stressor, the hypothalamus releases corticotropin-releasing hormone. This acts on the anterior pituitary gland to release adrenocorticotropic hormone into the bloodstream. Adrenocorticotropic hormone then targets the adrenal cortex, where it stimulates glucocorticoid synthesis and secretion. Cortisol, a glucocorticoid, then acts to regulate physiological changes in the body in response to the stressor. The chemical's influences don't end there, however. Cortisol is part of a negative feedback loop in the HPA axis. This means that when cortisol concentration in the blood reaches a certain threshold it should inhibit corticotropin-releasing hormone secretion. This suppresses adrenocorticotropic hormone secretion which then inhibits cortisol. This is one of many feedback loops that helps to keep the body in good balance. If it isn't happening there is some dysregulation occurring (Boileau, 2019).

Cortisol often has a bad name, but it does have some great effects. In the short term it can allow us to be very productive (Mason, 2018). It is released when the body goes into the 'fight-or-flight' state. When you think you are under threat cortisol enables you to have:

- a burst of energy;
- heightened memory functions;
- lower sensitivity to pain;
- elevated blood pressure.

These are all important to enable you to respond when you need to.

When people experience repeated or prolonged stress it is believed that the chronic activation of the HPA axis may lead to 'wear and tear'. Eventually this results in a dysregulated physiological stress response (Boileau, 2019). Regions that are involved in the stress response typically include the amygdala, frontal cortex and hippocampus. Chronically high levels have neurotoxic effects which can affect mood cognition and the stress response itself (Cloud, 2020).

> Glucocorticoids play an important role in mediating the remodelling of neurons in the hippocampus and frontal cortex during repeated stress exposure, likely due to regulatory effects on glutamate (Law, 2020).

An interesting study performed at Northwestern University looked at the cortisol levels throughout the day in older people (Adam et al, 2006). These people were also asked to note down whether they felt lonely, angry or tired. Normally when we wake up in the morning our cortisol levels are low, then about 30 minutes later we get a big boost. Throughout the day it reduces until it is at a low again around midnight (or three to five hours after the person goes to sleep). The results showed that those people who went to bed feeling lonely, sad, or overwhelmed got a bigger boost of cortisol the next morning than other people in the study who didn't feel these emotions. Emma Adam, one of the researchers, suggested this was an adaptive positive intention of cortisol to give these people the boost they needed to get up and out into the world so that they would increase their chances of connecting with other people.

This study also showed that angry people's cortisol levels remained higher throughout the whole day. From the other perspective, those who didn't get a big boost of cortisol in the morning felt tired throughout the day.

Liz Dunn, a social psychologist, did an interesting study that involved cortisol (Dunn et al, 2008). The team gave individuals $10 and told them that they could give as much of it away as they wanted, including nothing (keeping it all for themselves). As they had expected, they found that the more money people gave away the happier they felt. The more of the $10 they kept, the more shame they felt. The team was also able to measure the cortisol levels of the participants and found that the more shame people reported feeling, the higher their cortisol levels rose.

What are some scenarios where knowing about this would enable me to best serve my client or colleague?

In the working environment there are many occasions when cortisol levels increase. Any time that an individual's primary response might be to fight or flight it is likely that cortisol could be released. For example, if in a meeting they are being asked to justify their actions or thoughts; or when working with colleagues they find difficult. If they share that they are finding things troubling and their mood is lowered because of a certain situation, this could be an indicator. Obviously, we don't know for sure whether someone will have elevated cortisol levels, but there are some standard scenarios that make it more likely.

Caffeine is known to increase levels, as is sleep deprivation. So, any new parents are likely to be suffering a double dose! Severe calorie restriction has been linked to higher baseline cortisol levels. Travelling by train has also been linked to cortisol levels – the longer, more unpredictable and more effortful the train journey, the more cortisol is likely to be produced. So, when we ask people 'How was your journey?' we are likely getting some indicators as to their state, both in how they answer the question (we've all seen someone in a state of stress) and in what they answer. In an ideal world, getting those cortisol levels down before digging into work could be useful.

As a coach or manager, we might be able to see things that are triggering a stress response, but which our clients and colleagues cannot see themselves. Similarly, just because a trigger has not been identified does not mean they are not stressed.

It is also possible that a person may be susceptible to chronic stress. This could be due to genetic predisposition (genetic vulnerability) and/or physiological stress mechanisms and/or psychological processes such as stress appraisal, coping styles and perceived availability of resources (Boileau, 2019). We know early life experiences can affect vulnerability and resilience to stressors (Clow, 2020). It is thought childhood adversity is linked to blunted cortisol levels developing as an adaptation of the HPA axis to sustained periods of hyper-reactivity (Bunea, 2017). Atypical cortisol patterns (from any cause) are linked to psychological and physiological illnesses (Boileau, 2019). Flatter diurnal cortisol slopes (changes in levels from morning to evening) are associated with worse overall health. Challenges include immune and inflammatory system dysregulation, fatigue, cancer, obesity, and other mental and physical health challenges (Adam, 2017). Sustained high levels of cortisol have been linked to Alzheimer's disease, heart disease, insulin resistance, type 2 diabetes and have also been found to interfere with learning, memory, sleep, digestion, metabolism, impaired healing, mood swings... the list goes on (Mason, 2018).

What can I do with a client or colleague now having understood this?

Help each person identify what will work for them as their stress shower. How can they rinse (or scrub) off the stress of the day so acute stress doesn't descend into chronic stress.

Clients who understand what cortisol is – what effects it has on their body, when it is released and what they can do to decrease it – are empowered. The cortisol awakening response is considered a key link between mind and body. This is because we are sensitive to psychosocial factors. Taking the time to step into the teacher role as a coach and share some of the basics around cortisol can be useful. Equally can being fully present for your client, being non-judgemental and giving them the attention they deserve.

With links to both personal and professional challenges often clients can spiral deeper when they haven't realized what is happening. It is great to ensure that your clients have access to stress-reduction techniques or training. Thankfully, during the 2020 pandemic many organizations raised their game in terms of mental health training and support offered.

Mindfulness-based stress reduction continues to show promising results for the individuals involved in reducing stress levels. It is not yet conclusively agreed whether the practices influence cortisol levels (O'Leary, 2015). A meta-analysis of mindfulness-based stress reduction trials in 2011 called 'Mindfulness-based stress reduction and mindfulness-based cognitive therapy – a systematic review of randomized controlled trials' is a good reference source (Fjorback et al, 2011). There are lots of qualified teachers of this practice out there, who would make good guides and role models for those you work with if you are not yourself best placed. This practice offers lots of additional benefits too, which we focus on in Chapter 27.

Social relationships are very influential in cortisol activity and HPA responses. Couples in happier relationships have lower cortisol levels (Meyer, 2020). Social buffering is supportive in protecting people from the negative effects of psychosocial stress. Explore the social connectivity of the person you are working with. How are they connecting with others and how deep are these connections? Do they have between three and five close friends who they can call on and share important things with? Are they getting daily some connection with people?

Talk to your client about their physical activity levels. Regular physical activity – either in the form of an intensive session such as swimming, an aerobics class or running, or in smaller chunks such as jogging up the stairs or taking a power walk at lunchtime – can all lower the overall cortisol levels. These practices can also restore a blunted cortisol activation response to a healthy one in depressed individuals and in people who experience high stress (Smyth, 2020).

Consider asking your client to keep a laughter diary to record how often they laugh during a normal day. Dr William Fry has found links to laughter and lowered levels of stress hormones. Some people might benefit from watching a funny clip each day, reading a joke or, even better, meeting with friends and sharing things that make one another laugh. In a corporate environment consider the culture. Do people laugh? Do people have a sense of humour? Could small steps be taken to help people relax enough to find things funny?

The place of music in life is another consideration. Listening to relaxing music has been shown to lower cortisol levels. So, it could act both as an immediate support if a person is aware they are feeling in need of calming or listening a couple of times a day as maintenance.

Dopamine 11

Involved in how we think, behave, our feelings of motivation, reward and attention.

What is this brain chemical?

Dopamine is a neurotransmitter. It is a chemical that transmits signals between brain cells (neurons). It is one of the most heavily researched neurotransmitters.

Dopamine isn't made by lots of neurons, but the area of the brain called the substantia nigra is home to some dopamine-producing neurons. Another area is the ventral tegmental area (VTA). Dopamine acts through various distributed networks within the brain, but it isn't as widespread as some other systems, for example the serotonin system. It has an important functional role in the PFC and the basal ganglia. There are overlaps between the different networks that do have slightly different functions and innervate slightly different networks of brain regions.

Like all other neuromodulators, dopamine works in conjunction with other neuromodulators across distributed brain networks. Functions assigned to 'dopamine' and therefore only one part of a functional story which also includes other neurotransmitters. Dopamine containing cells are found in the raphe nucleus, which is traditionally a serotonin containing region. Dopamine is also a precursor to norepinephrine.

Perhaps dopamine is most well-known for its role in reward and motivation. Wolfram Schultz helped make dopamine's role in reward prediction errors well known. There are stacks of research on this. Now we know dopamine is involved in much more. However, there is still uncertainty about what dopamine's main function or functions are as we try to tie together all the different threads of research.

Why is it important to me as a coach/manager?

Dopamine is involved in:

- how we behave;
- how we think;
- our ability to move;
- our feelings of motivation;
- reward;
- punishment;
- working memory;
- attention;
- learning.

Reward prediction

Classic theories of dopamine focus on its role in 'reward prediction'. This underpins dopamine's role in learning. However, although the reward prediction signal is clearly found across many studies, there is still some debate about what these signals mean, and how they are translated into actions in humans. There is also a degree of heterogeneity in terms of how different pockets of dopamine neurons respond in relation to rewards.

Motivation

When something good and unexpected happens, such as being given a present, an ice cream, or praise in front of others, then dopamine neurons in a part (dorsal) of the VTA are activated. At this point you want to keep doing whatever triggered the release of dopamine. You feel motivated. When something bad happens, dopamine neurons in another part (ventral) of the VTA become very active. When you expect a reward, and it doesn't occur then dopamine neurons are depressed.

Learning

Dopamine is well known for its role in learning about rewards: learning from bad things stopping is like learning from a normal good reward.

Dopamine is also important in the frontal lobes for controlling the flow of information from other areas of the brain.

Instant gratification

At UCL, led by Professor Ray Dolan, it was shown that increased dopamine makes us more likely to go for something with instant gratification rather than a more beneficial, but slower to be delivered, reward (Dolan et al, 2010) – this obviously has key implications for coaches and managers. Participants were tested after being given a placebo, and after being given L-Dopa. The test involved making several choices between 'smaller, sooner' or 'larger, later'. For example, they could choose to receive £15 in two weeks' time or £57 in six months' time. Dr Alex Pine, who was involved in the study, said: 'Every day we are faced with decisions that offer either instant gratification or longer-term, but more significant reward. Do you buy your new iPhone today or wait six months till the price comes down? Do you diet or eat that delicious-looking cake? Do you get out your books to study for a future exam or watch some more TV?' (Dolan et al, 2010). The results of Dolan et al's experiment showed that each of the participants chose more of the 'smaller, sooner' options when they had more dopamine in their brains. This was great for the experimenters as they were only paying out the sum of £15 in two weeks rather than £57 in six months!

Decision making

University College London (UCL) is also responsible for a couple of interesting experiments investigating decision making. The first involved using imaging techniques to detect a signal in the brain that was linked to how much someone enjoyed an experience. It was then found that the signal could predict the choices a person made. The researchers suspected that this signal was dopamine, so they set up an experiment to see what happened if they tampered with the dopamine system (Sharot et al, 2012). A group of participants were given a list of 80 holiday destinations from all over the world to rate from one to six. They were given a sugar pill, to provide a control portion of the experiment, and asked to imagine themselves in half of the destinations (so individually in 40 of the original destinations). The participants were then given L-Dopa, which has the effect of increasing dopamine in the brain and is used typically to help sufferers of Parkinson's disease as it helps reduce tremors, rigidity, slowness and improve balance. These healthy individuals were then asked to imagine themselves in the other half of the destinations.

They rated all the destinations again, then the next day had to choose where they would go out of paired lists of holidays. The increased dopamine made people choose the holidays they had imagined when more of it was flowing through their brains, so they chose more from the second list.

The leader of the experiment, Dr Tali Sharot, said: 'Our results indicate that when we consider alternative options when making real-life decisions, dopamine has a role in signalling the expected pleasure from those possible future events. We then use that signal to make our choices.' So, higher levels of dopamine make us more likely to rate something favourably and subsequently choose it.

Mental health

The cognitive role of dopamine is evidenced by its implication in not only mental disorders, like attention deficit hyperactivity disorder (ADHD) but also healthy states of stress and fatigue, which are all accompanied by cognitive (and impulse) control deficits.

Cognitive performance

Dopamine is involved in cognition. This explains why some people have resorted to dopamine drugs (e.g. modafinil) to improve cognitive performance. The role of dopamine and the reward networks are explored further in the book 'Engaged; The neuroscience behind creating productive people in successful organizations'. Dopamine is involved in working memory, future planning component, intertemporal choice, selective attention, task switching and response inhibition.

Flexibility/cognitive control

One definition of cognitive control is that it is generally the mental processes that allow us to obtain our goals, by focusing on currently relevant representations and resisting distractions, temptations and impulses (Cools, 2019). Although traditional models of cognitive control focus on it being about making sure that mental resources are allocated effectively, other more recent theories pitch it in terms of a cost/benefit-based decision-making process. In other words, cognitive control is framed as a choice dilemma, is shaped by learning mechanisms that serve to maximize reward. Dopamine is implicated in cognitive control via its modulation of activity in the prefrontal cortex. When combined with dopamine's role in motivation, one proposal is that dopamine is involved in modulating our motivation to act and our cognitive control by modulating these cost/benefit computations, leading to selection of the best/most high value outcome based on more classical accounts of reinforcement learning – where iterative learning experiences lead to selection of the most valuable outcome/decision.

What are some scenarios where knowing about this would enable me to best serve my client or colleague?

Our experiences shape us. Dopamine receptors need more stimulation to trigger dopamine when we become addicted to something. In corporate environments it is often the norm for people to drink copious amounts of coffee. Constant caffeine, along with saturated fats and refined foods, are not great for brain function. When we eat a lot of junk food our dopamine production is inhibited. This can have a direct impact on our productivity. As time goes by, we can need more caffeine to get that same 'hit' – and this negative spiral is not ideal. The same applies for people using tech to constantly stimulate themselves. This has been recognized most publicly in Silicon Valley, where doing 'dopamine fasts' have become fashionable. While the science on this approach isn't 100 per cent, fasting is an amazing discipline and various types of fast offer numerous benefits.

Organizations often tell us that things need to be gamified for their people. Could we sort out the mental health problems that are so widespread with a five-minute video? (Because they have short attention spans these days!). While we do indeed value the bite-sized resources that can be very impactful... we also caution organizations. You have a problem if your people can't pay attention for 30 minutes to something. Or submerge themselves in deep work. Attention is a precious resource, dopamine is clearly involved, but continuing to ask for less and less in terms of focus and attention isn't the answer.

Many biases are influenced by various neuromodulators. Biases involving not being able to delay gratification are linked to dopamine. It is likely biases linked to attention, salience and information are also. Retraining responses to biases is possible but is often poorly achieved through programmes. Breaking down the underpinnings of a bias is key. The key message here is not to give up on people. They can change... if the first strategy didn't work consider a new one.

What can I do with a client or colleague now having understood this?

People's experiences, environment and thoughts all influence their dopamine levels and receptors. Remember that more isn't always better. It depends where you are starting from, and what you want to do. Teaching your clients how to trigger a dopamine boost from just thinking about the small tasks that are in alignment with their goals will make it much easier for them to follow through with their decisions. For example, when we think about cake many of us experience a dopamine release, which will motivate us to seek out cake! Eating the cake will also trigger a release, programming us to want it again. We've just learnt that having cake makes us feel good. For most people, this isn't ideal on a regular basis. As with any long-term goal you need to be mindful of your short-term actions. This is where thinking about how you'll feel after you achieve something, rather than necessarily how you'll feel actually doing the tasks, can be useful. In essence we need the 'smaller, sooner' rewards to come from things that are in alignment with the 'larger, later' goals. This will likely take an investment of time and energy. This is a key benefit the coaching process can offer. Time to step back and check alignment between weekly actions and long-term goals. A lot of our work with managers involves equipping them to have these conversations that are not all so transactional. Giving team members the opportunity to step back and consider how things are set up to trigger specific actions.

Consider our daily routines. When we think about clearing some e-mails many of us experience that same dopamine release; this helps us want to just sort through a few, perhaps replying with a shorter reply than it deserves just so we can delete or file it. Once we reply to one, we often want to send another and another in order to feel that mini buzz when we click 'delete' – so that it is no longer in our inbox and lobbying for our attention. We often find ourselves 'doing e-mails' when there are other things that we could more beneficially give our attention to.

Simply being aware of how easily influenced we are by our neurochemicals is useful to those you work with. Individuals need to get to know themselves, and managers their team members. Good coaching conversations can help people to realize what gives them satisfaction. How they can replicate the experiences that fuel and drive them. The old advice to managers of getting to know your team couldn't be truer here. What is engaging and compelling for one person could be really depleting for another. The more self-aware your people or your clients are the better you can support in the creation of High-Performing Neural Environments for them and with them.

Oxytocin 12

Involved in social behaviour, perhaps best considered a social magnifier.

What is this brain chemical?

Oxytocin is an evolutionary ancient neuromodulator and hormone. It is primarily produced in the hypothalamus from which it is secreted both within the brain and into the circulatory system. Behavioural brain-based effects of oxytocin include modulation of maternal behaviour, infant behaviour, social bonding, sexual and agnostic behaviour, group membership, feeding, stress and anxiety. Oxytocin represents a hormone that promotes social cognition and mentalizing. It can promote the formation and maintenance of affiliate cooperative relationships in diverse situations. It may do this by increasing sensitivity to social rewards and social threats. It can help us with cognitive and affective social information processing. This means the complex impacts of oxytocin can be surmised to be positive, negative, or aggressive (Maroun, 2016). This is in stark contrast to its current widespread reputation as the love hormone, moral molecule, or cuddle chemical.

Oxytocin has the effect of:

- reducing blood pressure;
- reducing cortisol levels;
- increasing pain thresholds;
- an anti-anxiety effect.

It used to be thought that oxytocin was a hormone with important effects only for women around childbirth and breastfeeding. Now the additional important and wide-ranging effects in social processing are more recognized.

If we had to give a one-line summary for oxytocin at this stage of the research, though, I'd favour the one from Larry Young: 'Oxytocin is tuning us into social information and allowing us to analyse it at higher resolution' (Young, 2014).

Why is it important to me as a coach/ manager?

Pro-social or anti-social?

The unsatisfactory answer people dread is… it depends. But the evidence is very clear now. It does depend on various factors. We cannot label oxytocin as a molecule that exclusively enhances pro-social behaviour. It absolutely does have pro-social influences. However, in some situations it may enhance anti-social behaviour (Quintana, 2020). When you think of this logically, it makes sense. Imagine a scenario where a mother needs to defend her young child. A molecule that helps her transition from peaceful, caring nurturer into feisty defender is valuable (Crespi, 2015). In an organizational context this could extend to defending your team, especially if there are strong bonds or you feel a sense of unfairness. Oxytocin is believed to be able to dampen the cortico-amygdala circuitry thus providing alternative responses to threat and may help us regulate vigilance and defence-motivated aggression aimed at self or group protection (De Dreu, 2015). Oxytocin could be implicated when people lie and cheat in order to benefit their 'in-group'.

Mental health

Oxytocin has been implicated in psychiatric disorders such as depression, anxiety disorders and post-traumatic stress disorder (Matsushita, 2019). In part due to its involvement in social emotional behaviour and role in healthy long lasting social relationships oxytocin has a pivotal role in good mental health (Hurlemann, 2016). The research is ongoing in this space. There is a suggested possible relationship between peripheral oxytocin levels, oxytocin receptor gene polymorphism, and generalized anxiety disorder (Newman, 2016). There are also clinical trials exploring oxytocin's efficiency in improving social deficits in a variety of conditions with associated social dysfunction. This could impact disorders such as autism, schizophrenia, depression, social anxiety and drug addiction (Erdozain, 2020). It is important to remember most research is done on mentally healthy individuals unless otherwise stated. The role of various neuromodulators may be different in people with mental health challenges. For example, with those diagnosed with borderline personality disorder oxytocin has been found to decrease trust and cooperation (Piva, 2018).

Trust

The effect of oxytocin on trust seems to be much less robust than earlier studies suggested. It seems now… as it does with many of the other themes… that the effect of oxytocin on trust is more complex than a simple 'it increases it'. Other factors need to be taken into consideration, such as attachment, personality traits and situational factors (Ebert, 2017).

Further studies

Neuroscientists Michael Steinman and Brian Trainor have further developed the idea that oxytocin may have different effects in men and women. Their work also suggests that under certain circumstances the hormone may even trigger anxiety. Other recent studies have shown that it can intensify a negative memory of a social experience or increase the likelihood of aggression and violence towards people who are not part of your social group (Steinman, 2015).

A study from the Centre for Music in the Brain showed that when subjects received oxytocin they were more synchronized in their finger-tapping. The experiment was done with pairs of participants placed in separate rooms, rapping, in a leader–follower relationship. Some were given oxytocin and others a placebo. Oxytocin can be involved in social connectedness (Gebauer, 2016).

In a completely different study at the University of California, Irvine found that oxytocin reinforces social ties by inducing anandamide formation in the nucleus accumbens. Anandamide is sometimes called the 'bliss molecule' because it activates cannabinoid receptors (like marijuana) and heightens motivation and happiness (UCI News, 2015).

What are some scenarios where knowing about this would enable me to best serve my client or colleague?

It's complicated.

Not the headline any coach or manager really wants to see. However, oxytocin's effect on social behaviours is far more complex than we initially thought. They include the enhancement of negative social emotions as well as the well-publicized positive ones. It has become clear that oxytocin is not likely to be the miracle molecule that works in a 'one size fits all' way the research once suggested (Erdozain, 2020).

Consider your context

It is no longer accurate to say, 'raise oxytocin levels to do x or y'. You need to consider the context you are working within. If you are managing a team and one member of the team is clearly behaving in an untrustworthy way or isn't yet known to the rest of the team, then the positive effects oxytocin can have on trust are not likely within this context (Piva, 2018).

Social seekers

We are social creatures. As you will read later, loneliness is a killer. Managers need to be aware of just how important social interactions are for their people. Workflow needs to be designed with this awareness in mind. Physical and virtual environments need to respect this reality. Many people will seek out their own social moments if the organization isn't predisposed to facilitating them. The richer you can make these social interactions, as a manager, the stronger team you'll have. If working independently as a coach, we need to remember clients need these pillars of social cohesion in the different walks of their life.

Vulnerability and openness

There are so many situations where people would benefit from being a bit more open and a bit more vulnerable. Managers are so influential in the performance and lives of their teams. We've seen such benefits when real relationships are strengthened at work. One of the simplest ways to do this can be to just open up a bit. If you want a coaching relationship (either external to work or internal) to be strong then trust is vital. If we're expecting a client or colleague to open up and be vulnerable this process can be fast-tracked if you are willing to go there first!

What can I do with a client or colleague now having understood this?

Increasing social acuity

While everyone is born with genetic predispositions... there are also so many opportunities for tapping into your potential. When it comes to the oxytocin system we're thinking about your social skills. Isolating and practising the various micro skills that add up to create a socially flexible and competent individual is a worthwhile investment. Research shows many social skills involve oxytocin. It would be logical that just as oxytocin influences social skill-based responses to situations, the practising of particular social skills would influence the oxytocin system.

Capitalizing on experiences

Being aware that intense experiences, be those that evoke fear or excitement or surprise, have the potential to increase group cohesiveness. Consider how you can make the most of these opportunities either through coaching an individual or as a manager of a team. Going on a rollercoaster with your boss or doing a skydive with a new partner can be pivotal experiences, with potentially long memory traces.

Unhelpful sources?

Some people may be seeking to boost their social connection with others through unhelpful means. Many people have had negative experiences with the swipe left swipe right approach to building a relationship. When a deep connection is being sought then a quick surface level interaction can often leave people feeling worse. Perhaps you know people who have jumped on a social media platform, perhaps not even knowing why, but maybe just to feel a little connection.

Serotonin   13

Important for mood regulation, appetite, sleep, memory and learning.

What is this brain chemical?

Serotonin is one of the most widespread chemical modulators in the brain. It's involved in many functions. In fact, it's almost hard to find functions it's not involved in. Although there are only 300,000 serotonin cells in the human brain its projections are so widespread that there were a few cells in the brain that aren't under the influence of serotonin in one way or another. It is highly multifunctional. If it had feelings, it might be offended by so frequently being reduced to 'the happy hormone' by so many.

Serotonin is also known as 5-hydroxytryptamine (5-HT), which becomes more interesting when you know that serotonin is made from tryptophan because we can see the link between one of the precursors and the end product. It is one of the most well-known neurotransmitters and has its reputation firmly established for being important to feeling happy (through its link to depression). However, serotonin is implicated in several mental illnesses. Studies have shown how abnormalities in the serotonin system, for example in terms of the expression patterns of specific serotonin receptors, are associated with disorders such as autism spectrum disorder, schizophrenia and anxiety. It has also been implicated in disorders characterized by aggressive behaviours.

Tryptophan, the essential amino acid needed to make serotonin, is found in our diet. Bananas, dates, yoghurt, milk, chocolate, sesame, sunflower and pumpkin seeds, and poultry all have high levels of tryptophan. Some foods are extra special because they have a high ratio of tryptophan to phenylalanine and leucine (two other amino acids) and this ratio can influence serotonin levels. So, eating dates, papayas and bananas could be a good plan if tryptophan levels might be low.

While serotonin is released by neurons in the pineal gland of the brain it has effects all over the body. It is involved in regulating the sleep cycle and circadian rhythms (along with melatonin) and is affected by sunlight. Sunlight has ultraviolet light as a constituent part and our skin absorbs this to produce Vitamin D, which is important in promoting the production of serotonin. The exposure to sunlight is something many people find more challenging in the autumn and winter months. The rise of seasonal affective disorder (or SAD) is one of the results we see. Either getting more sunlight or using a well-researched, credible light box can help boost serotonin levels and will hopefully improve mood.

When we say that people have low serotonin activity or levels, what is actually happening inside the brain can be one of several things. One explanation is that your brain is making less serotonin. Another is that there are fewer serotonin receptors or that the receptors there are not receiving the serotonin well. Alternatively it might be that although you are making lots of serotonin it is being reabsorbed too quickly back into the neuron from the synapse, or that it is being broken down too soon.

Why is it important to me as a coach/ manager?

Mental health

Some of the figures state that 1 in 10 Americans are taking an antidepressant. In some cities in the United Kingdom, 1 adult in 6 is on antidepressants (Hutchison, 2011). One of the most popular forms of antidepressants is what is known as a selective serotonin reuptake inhibitor, or SSRI. These work by increasing the amount of serotonin in the synaptic gap by inhibiting the reuptake of serotonin back into the presynaptic cell. This means that there is more available to bind to the postsynaptic receptors. Recent studies have looked at the effectiveness of SSRIs in people who are mildly, moderately or severely depressed. People suffering from mild and moderate depression, it was found, received small or no additional benefit when compared to taking a placebo. The results are less conclusive in people with severe depression.

Wellbeing

Serotonin and dopamine work together to help regulate the appetite. When we smell food, dopamine is released to increase our appetite. While we are then eating food, serotonin activates a type of receptor that affects dopamine-producing cells. The effect is to stop the release of dopamine, thus reducing appetite. In the absence of serotonin, one is less able to identify when they are no longer hungry, and this is linked to weight gain. Serotonin levels can be raised. Getting enough sleep is important. Coaches should be wary of being too prescriptive on this point, though. What constitutes 'enough' sleep for one person can be very different to what is 'enough' for another person. A change is normally a good indicator of something worth addressing. If a person had been sleeping well for eight hours and recently has been waking up after five, unable to get back to sleep, then it is possible that this could be affecting them.

Massage

Some research has been done into massage and serotonin levels. A study was carried out with depressed pregnant woman (Field et al, 2004). These subjects received two massages per week from their partners for four months. A 30 per cent increase in serotonin levels was measured. Another study looked at the effect of massage on babies. These babies had depressed mothers, so this was another group where this research made a lot of sense. These one- to three-month-olds were massaged for 15 minutes for six weeks and their serotonin levels increased by 34 per cent. There are other studies that also explore a non-loved-one doing the massaging. Overall, looking at several studies we could generalize to say that massage appears to decrease cortisol levels and increase serotonin and dopamine levels. This has a positive effect on people with a variety of medical conditions and experiencing stressful things.

Stress

The serotonin system has also been implicated in stress reactivity. Carhart-Harris and colleagues suggest a flexible coping model of brain serotonin function. This includes the idea that a key function of brain 5-HT is to moderate anxiety and stress and promote patience in coping. They also propose a second major function of brain 5-HT is to open a window of plasticity for greater adaptation. It is exciting that serotonin could mediate stress moderation and plasticity mediated adaptability in response to different levels of stress and adversity.

Cognitive vulnerability

There are multiple ways to interpret or evaluate a situation. Similarly, there are multiple ways to remember the past and imagine the future within your own thoughts. Evidence suggests that serotonin may have an influence on these processes, therefore influencing whether people perceive/attend to/remember what's positive vs what's more negative and having the potential to increase cognitive vulnerability to negative or aversive events or situations. For example, evidence suggests that some people who carry a particular variant of a serotonin receptor are more prone to ruminations such as negative attributional styles and self-referent memory biases while studies of serotonin effects on facial perception of healthy volunteers report that differences in serotonin availability alters identification of fearful faces. Genetic differences relevant to serotonin expression also impacted recognition of facial expression of emotions. These changes are proposed by some to underlie the link between depression and serotonin (the so called cognitive-emotion model of depression).

Managers, how are you supporting your teams' mental wellbeing? What specifically do you do each week that builds their resilience? Does your client or colleague have manageable workload levels? Is their life so full that they must sleep less to get everything done?

What are some scenarios where knowing about this would enable me to best serve my client or colleague?

You know that the mind is very powerful. Even thinking about things has a profound impact on our biochemistry. Although you know this, not everyone does yet. Calling to mind something positive – such as a happy memory, an achievement, a great experience – increases serotonin production in the anterior cingulate cortex (ACC). This is an area of the brain just behind the prefrontal cortex that controls attention. On the other hand, if clients find themselves ruminating on sad events then they can expect a decrease in serotonin production in the ACC. Being consciously aware of what is filling your mind and the effect it has on you is important and can dramatically affect how your day goes. Systemizing your thoughts may sound a little drastic but can be very useful in creating useful habits and getting your biochemistry to a place where you can then be on more of a roll. For example, when coaches suggest starting the day with an 'attitude of gratitude'-type exercise this could set people up for a more productive day. Managers can influence where their people's attention goes on a daily basis. What people are focusing on influences their HPNE. Knowing a person's current predisposition shouldn't fix them into a box.

Exercise has been shown to be important both in serotonin production and release. One of the most common things managers we work with during Synaptic Potential's Management Development Programmes tell us is that they would love to prioritize their own exercise each week. So, for anyone who wants more motivation to start exercising, or to continue exercising, this knowledge might be helpful. Typically, the research focuses on aerobic exercises, which are great for boosting serotonin. However, if your person is a yoga fan then that could work too. There appears to be a difference between choosing to exercise willingly, and exercise that you feel forced into doing. A point for another time, though, is that often to create a new helpful habit you may have to do things that sometimes you're not feeling the love for, but it will come later. If a client is struggling to exercise willpower to delay gratification in some way then their serotonin activity may be low. It may also be proving difficult for them to create and deliver on well-formed plans. When we consider what is normally necessary to achieve a goal then we can see that low serotonin activity may make this more difficult. Sometimes there are steps that need to be taken before a goal is worked on. Building up an individual to a place where they stand a good chance of achieving a goal can be a really good exercise.

What can I do with a client or colleague now having understood this?

Overall, anyone looking to work on a goal should do well to reflect on times when they achieved their goals. Not only is this in itself likely to release serotonin, there may also be things they could learn to help them get into optimal goal-achieving states! In a corporate environment, however, it could be useful to look at the ways in which it is easy for people to raise their serotonin levels. It is important here to remember that sunlight, massage, exercise and memories of happy events are all great ways to boost serotonin levels. You might want to explore how they as an individual can take personal responsibility for how they get these activities into their life more. If someone is looking to increase the serotonin levels of others who they work with then they could find creative ways to encourage these activities. Perhaps a team exercise competition, where over the course of a month participants post publicly how often they are exercising. The people who stick to their goals win a massage! Could any meetings be held outside rather than indoors in order to boost sunlight exposure? How do happy, successful events get remembered corporately? These are some of the questions and thoughts you might be having.

When working with individuals in a personal capacity you have even more scope to hold them accountable to their own levels of serotonin and happiness. As mentioned above, sometimes taking the time to work on preparing someone to undertake a goal is the best way to start. We run sessions on 'Change Readiness Neural Quotient'. There are evidence-based ways to prepare people for change and new habits. A good place to start is to coach a person using lots of open questions to help them identify how they think they can get their serotonin flowing, and cross-referencing this with what the science tells us. Some people know themselves really well, and so trusting what they intuitively feel would be good strengthening exercises for them could also work well.

Noradrenaline   14

Best known for its role in the fight-or-flight response – but also involved in lots more.

What is this brain chemical?

Noradrenaline is also known as norepinephrine and in the future, it is likely everyone will favour the term norepinephrine. It is both a hormone and a neurotransmitter. You have probably heard about adrenaline (only slightly different chemical structure) and its role in the fight-or-flight response. It is important for regulating our heart rate, blood vessel and air passage diameters and has metabolic impacts.

Adrenaline acts on nearly every tissue in the body. In the heart it increases the rate at which it beats. It can cause both vasoconstriction and vasodilation (constricting and dilating the muscular walls of vessels). In the lungs it increases the rate at which we breathe. In the liver it stimulates glycogenolysis, which means the breakdown of stored glycogen to glucose and glycogen. It triggers lipolysis, meaning the breakdown of lipids into glycerol and fatty acids, giving us fuel to use. It also can cause muscle contraction.

The classic way that people remember the basic effects of adrenaline are to consider what we want to be able to do if a lion appeared in front of us. By our heart pumping faster there is more blood with all its available nutrients being sent quicker around our body. Combine this with us taking in more oxygen by breathing at an increased rate and this means that we are primed for action. Add in the action of breaking down glycogen to make more glucose available and we're ready to run.

Noradrenaline is a direct derivative of dopamine (Wahis, 2021). Noradrenaline and dopamine are required for proper prefrontal cortex functioning. An impairment in each leads to disrupted working memory (Xing, 2016). It is believed the locus coeruleus is an important hub for broadcasting command signals from the PFC to other brain areas (McBurney-Lin, 2019). The LC-NA system is well positioned to influence global brain states. The NA appears to be released from synaptic bulbs and to diffuse beyond the bounds of a typical synapse to interact with adrenergic receptors on surrounding neurons and glial cells (Ross, 2021). Noradrenaline is involved in a range of central nervous system functions such as waking, arousal, attention, sensory discrimination and cognition (Borodovitsyna, 2018). It promotes PFC functions such as working memory and top-down attention mechanisms. Also involved in decision making and emotional regulation.

Why is it important to me as a coach/ manager?

Noradrenaline is involved in our emotional response to things. This process is controlled by the sympathetic nervous system. Fear is the best studied of the emotions in relation to noradrenaline. Studies have shown that when we have higher levels of noradrenaline we show more negative facial expressions in response to a scary film, for example, than people with lower levels of noradrenaline. The studies also reported a greater intensity of negative memories. The same was not the case for other emotions. So, people given a noradrenaline boost did not find funny films funnier, for example, or angry films angrier.

Noradrenaline can produce retrograde enhancement of long-term memory. When an emotionally stressful event occurs, adrenaline is released. This can modulate memory consolidation. If we think about this, it is very useful in theory for us to better remember the things that we are marking as more important than those that are less important. In practice there can be some complications with this, however. Some of the complications include post-traumatic stress disorder, where noradrenaline is thought to play a role in increasing arousal and fear memory. It is also implicated in attention-deficit hyperactivity disorder.

An interesting effect of noradrenaline is 'broken heart syndrome' or Takotsubo cardiomyopathy. This syndrome affects people who have suffered severe emotional or physical stress. Often it is seen in older women after their husbands pass away. Initially the symptoms resemble a heart attack but a full recovery is normally made within days or weeks. The process that researchers believe is occurring is that the body changes its response to noradrenaline by switching from stimulating the heart to reducing its pumping power. This is important for medics to be aware of, so they do not give more noradrenaline. It is also interesting to us that the heart may change its response to adrenaline, hypothetically to protect it from being overstimulated from high levels of noradrenaline that are normally released during stress.

Until recently we have associated chronic stress with damage to DNA. In the last few years, a mechanism has been proposed that helps to explain how this happens. When people suffer from chronic stress they have elevated noradrenaline levels. Researchers found that chronic stress lowered p53 levels. This is a tumour suppressor protein, and its role extends to preventing abnormalities of the genome. In mice this effect was noticed in just four weeks.

On a more positive note there are lots of great times when noradrenaline is also involved. When we are falling in love our body releases a mix of chemicals, including adrenaline, noradrenaline and dopamine. The dopamine may make us feel euphoric. Adrenaline and noradrenaline make our hearts beat faster and make us feel a bit restless and preoccupied.

Knowing the links between chronic stress, high levels of noradrenaline and long-term potential health problems are important. It establishes the stakes that people you work with may be gambling with.

What are some scenarios where knowing about this would enable me to best serve my client or colleague?

Attentional states and our ability to control our attention involves regulation by the LC-NA system. Someone struggling with attentional challenges may be experiencing dysregulation of the LC-NA system. Without the ability to maintain optimal levels of NA it can be hard to control attention in a goal-directed manner (Unsworth, 2017). Unfortunately, a single, chronic stressful event can alter LC-NA function. Early adolescence is a critical window when social stress can strongly affect behaviour and brain NA activity (Xing, 2016). Exposure to chronic stress is able to alter LC gene expression, promote long-term changes in synaptic transmission and excitability and induce morphological changes and dendritic remodelling. The brain is being rewired. These changes can influence changes elsewhere and ultimately can cause chronic anxiety-like behaviour (Borodovitsyna, 2018).

Imagine you are working with a senior leader in an organization. You have been working with them for some months now. For as long as you have known them they have been suffering from chronic stress. They have raised this on several occasions and each time said that they would deal with it later. You are starting to notice a pattern. By understanding the high stakes that they may be playing with, you are in a position to share this with them. You might invite them to consider the research and whether they are still happy playing on this high-stakes table.

What can I do with a client or colleague now having understood this?

In a work environment – so for corporate coaches and managers in the role of coaches – doing a noradrenaline audit could be useful. Looking at an average week, what are the times that stand out where a noradrenaline surge may be occurring? For some people this will be obvious, and they will be very connected to the feelings they experienced linked to a noradrenaline surge. At this point the emphasis is on awareness for personal management. So, by becoming aware of when, and how often, a big noradrenaline rush occurs, people can then consider whether they want to do anything about that.

In some lines of work a regular noradrenaline rush is considered normal and individuals are not experiencing any adverse effects from it. Perhaps for people in sales, the thrill of the chase and being on tenterhooks before finding out whether a proposal has been accepted or not is all part of a healthy process. For other people, tense relationships, poor management and unhelpful systems can all lead to noradrenaline bursts that negatively affect health and may warrant investigating.

For many people, noradrenaline is only being released appropriately and is not linked to any chronic stress problems. As coaches and managers we have to keep our curious hats on and avoid presuming. Trusting our clients and colleagues to work things through for themselves can sometimes be a cause of noradrenaline release in us!

GABA
and glutamate

15

The king and queen of neurotransmitters.

What are these brain chemicals?

Glutamate and GABA are very important neurotransmitters. They are the most fundamental of all the neurotransmitters. For coaches they are a little less easy to position as sexy and relevant but they are included here because they are so key to brain function. The biochemistry that underpins their activity is understandably quite full and there is not space, nor probably the interest from most, unfortunately, to warrant its inclusion. So here is a simple overview of what they are.

Glutamic acid is a non-essential amino acid, meaning that it can be made in the body from essential acids. Officially it is the carboxylate anions and salts of glutamic acid that are known as glutamates. Glutamate has very important roles as a neurotransmitter vital for long-term potentiation (a form of neuroplasticity) and involved in learning and memory. Glutamate is the most abundant excitatory neurotransmitter in humans.

Interestingly, most of us are familiar with the manufactured salt flavour of glutamate in the monosodium glutamate used in a lot of Chinese takeaway foods. Glutamic acid is responsible for the flavour we call umami, one of our five basic tastes.

GABA is gamma-aminobutyric acid, the chief inhibitory neurotransmitter (just as glutamate is the chief excitatory neurotransmitter). It acts to inhibit signal transmissions between nerve cells. Glutamine availability impacts the manufacture of GABA. Adequate levels of GABA impacts peace of mind, balance, concentration and restorative sleep.

Why are they important to me as a coach/ manager?

Glutamate's involvement in learning and attention is relevant to coaches. Several studies have shown that chronic stress can trigger a decrease in glutamine levels and a suppression of its function in the prefrontal cortex (PFC). This can have a detrimental effect on PFC-dependent cognitive processes, of which we know there are many. This means that there is a very real scientific underpinning to chronic stress, making it more difficult to think, process information, make decisions and all the other cognitive processes that many of our clients need to perform every day.

During acute stress episodes, cortisol is secreted from the adrenal glands. Certain areas of the brain (hippocampus and PFC) have a lot of cortisol receptors. When they are activated an increase in the amount of glutamate is noted. More glutamate in the short term leads to heightened performance in these brain areas, as part of the fight-or-flight response. The glutamate is normally cleared from the synapse but if there isn't enough time between stress phases, as is the case with chronic stress, then the high levels of glutamate can lead to excitotoxicity; damage to cells causing a reduction in brain areas.

A few other interesting facts include that the more glutamate receptors someone has the higher their intelligence. 'Overexcitability' from glutamate can be controlled by diet reductions in gluten, casein, glutamate (eg monosodium glutamate, or MSG) and aspartate. Alcohol abuse is linked to higher levels of GABA, and higher levels of GABA have been linked to depression.

PART THREE
Foundational brain concepts

In Part Three we focus on the foundational things that a coach and manager need to understand about the brain and what this means to them. We look at the core units of the brain, the neurons and synapses. Then we explore the fascinating area of neuroplasticity. When we teach neuroplasticity in depth in the Neuroscience for Coaching programme, people often tell us that they find it really exciting. They recognize that the area of neuroplasticity is fundamental to coaching and it offers a scientific explanation for a lot of what they experience in their coaching and people development work.

There are several conceptually useful models that didn't make it into this section because they are scientifically disputed and, while it is possible that others may need revising in future years, we have a good balanced snapshot between usefulness and science today.

The threat response is of particular importance in a work setting. The problems that arise from the threat response occurring in individuals, and this not being recognized, frequently lead to mismanagement. Lower efficiency, effectiveness and productivity are all side effects of the threat response. Being aware of it gives people the opportunity to act differently.

We also investigate the various types of brain scans that tend to be mentioned in studies that you may be interested in and how they work. This overview is very useful to anyone wanting a deeper understanding of what is being discovered in the field.

The working memory is the fifth topic in this section and is a good 'general awareness' area to have up your sleeve. With some concepts it is not likely that you would discuss them very often with people, but being aware of them may link into your work in other ways. The hypothalamic-pituitary-adrenal

(HPA) axis, by contrast, is something that is being activated in those you work with on a regular basis and so an understanding of it may be of use frequently.

The underpinnings covered in this section provide a firm foundation for the following parts of the book, in which we look at diverse constructs such as trust, willpower and powerful influencers on behaviour.

Neurons and synapses

What are they?

The body is made up of cells. Neurons are otherwise known as 'brain cells'. They are electrically excitable cells that process and transmit information. They communicate either electrically or, more commonly, chemically. Neurons have various components to them including a core cell body with a nucleus (like the cell's mini brain), an axon (a long single extension coming out of the cell body), and dendrites (thin structures that also extend from the cell body, but of which there are many and branch multiple times giving the illusion of a 'dendritic tree'). The axon is insulated by a myelinated sheath, which enables electric currents to pass efficiently down the axon. At the end of an axon there is a little bulb, the axon terminal (sometimes called the synaptic knob).

The place where one neuron communicates or transmits a chemical signal to another is called the synapse. They are the specialized connections through which cells communicate chemically. Although we call them a connection, in fact the synapse is also called the synaptic gap. There is actually a gap between one synaptic knob and another dendrite. We can think of this first part, the presynaptic neuron, as having something to communicate and it is the postsynaptic cell that is prepared to listen.

An action potential (a nerve impulse) travels down the axon. When it reaches the axon terminal synaptic vesicles containing neurotransmitters are released. These neurotransmitters then travel across the synaptic gap and attach to receptor sites on the postsynaptic cell. This may or may not generate another action potential here. There is no other option, so it isn't a question of strength of action potential – it is either a stop or go situation, a yes or no, an action potential or no action potential situation. Each neuron can be connected to as many as 15,000 other neurons through these synapses.

Receptor sites that are positioned on the postsynaptic cell are ready for neurotransmitter molecules to bind with them in a lock-and-key-type fashion. When this happens they are telling the cell to do something. Each receptor is specific and will only bind with a specific neurotransmitter. The action is either excitatory or inhibitory, so it either results in the generation of an action potential or it does not.

The process that occurs at the synapse does not offer direct applications for a coach; however, a general understanding of the process is a good foundation and is as follows:

1 The neurotransmitter is made, or synthesized. This can happen in either the cell body, the axon or the axon terminal.

2 The neurotransmitter is stored in synaptic vesicles in the axon terminal, waiting.

3 Calcium ions enter the axon terminal during an action potential (a nerve impulse) and this signals to the synaptic vesicles to release their neurotransmitters into the synaptic gap.

4 Once the neurotransmitter is in the synaptic gap it binds to and activates the receptor sites in the postsynaptic membrane (the edge of the postsynaptic cell).

5 Finally, the neurotransmitter is deactivated. Either this happens by it being destroyed by enzymes or it gets taken back into the presynaptic neuron to be reused or destroyed.

For videos that bring this process to life check out:
www.neuroscienceforcoaches.com

Why are they important to me as a coach/ manager?

This fundamental process is very important. The formation of memories at its most basic level occurs at the synapse. As neurotransmitters communicate from the presynaptic neuron, across the synaptic gap to the postsynaptic neuron, the connection between the two neurons is strengthened. The analogy we use during the Neuroscience for Coaches training programme and the Coach Savvy Manager training programme is that it is like two people passing a ball between themselves. Each person is a neuron and the ball is the neurotransmitter. Normally the passing of the ball is a little slow and clunky, sometimes it even gets dropped on the floor. However, the more they throw it back and forth, the more the neurons communicate with one another, the quicker they are able to. We talk about the connection between the neurons as being strengthened and the strength results in the storage of information, which results in memory. This process is called long-term potentiation.

Long-term potentiation is described as a long-lasting enhancement in the transmission of signals between two neurons as a result of stimulating them synchronously. It is so important for coaches because it underlies synaptic plasticity, or neuroplasticity as it is more commonly known. This ability for synapses to change their strength is vital for learning and memory.

This process of learning and changing is vital for anyone involved in people's development to understand. It influences change programmes, learning design, habit formation and more. When you understand what is going on in people's heads you can design environments and processes that lead to the desired results more easily. Regularly creating a High-Performing Neural Environment relies on our ability to strengthen new neural connections in the brain.

Neuroplasticity 17

What is it?

Neuroplasticity is the property of the brain to change. There are several different ways that the brain naturally changes. When the strength of the connections between synapses changes we call this synaptic plasticity. This focuses our attention on the location of the change. A little more specifically, long-term potentiation could be occurring (we mention this in Chapter 8 on the hippocampus and Chapter 15 on GABA). Alternatively there could be more or fewer receptors present for specific neurotransmitters, or a number of other things could be different. Action at the synapses is very important within the brain.

Other changes that come under the broad banner of neuroplasticity include synaptogenesis and synaptic pruning. This is the effect of creating or removing the whole synapse, adding or taking away connections between neurons. Neurogenesis is the creation of whole new neurons, and neural cell death is when neurons die (this can be a natural process, but could also be as a result of damage, overexcitation or disease). These usually come from brain scans and can involve changes in the density of white matter or grey matter on structural MRI scans, or how densely radioactively labelled markers bind to specific receptors in parts of the brain.

There are other forms of neuroplasticity, including when we see from brain scans that areas of white or grey matter have become more dense, but we don't actually measure what individual neurons are doing. We look at examples of this in Chapter 27 on mindfulness.

As is common with things relating to the brain, we learn a lot when things go wrong. Brain injury has given us a good insight into the necessity for neuroplasticity. If an area that we believed was solely responsible for movement of the left arm, for example, was destroyed in an accident, then the old way of thinking would suggest that the person would not be able to move this arm again. However, with neuroplasticity it is possible for cortical remapping to occur, so for the role of moving the left arm to be taken over by other neurons in a different place in the brain.

This isn't the only time that the brain changes. In fact it turns out to be a normal process that is happening regularly. Behaviour changes, environmental changes and thoughts can all effect change in neural pathways and synapses. You'll remember when we looked at the synaptic gap and neurotransmitters binding to the receptors on the postsynaptic cell that either an action potential was generated or it wasn't. The additional piece of information here is that the *speed* and therefore the *frequency* of the action potentials can be influenced. Typically, the more this particular pathway is activated, the faster and so more frequently a nerve impulse can be seen to be travelling.

A famous paragraph written by Donald Hebb in 1949 reads:

Let us assume that the persistence or repetition of a reverberatory activity (or 'trace') tends to induce lasting cellular changes that add to its stability... When an axon of cell A is near enough to excite a cell B and repeatedly or persistently takes part in firing it, some growth process or metabolic change takes place in one or both cells such that A's efficiency, as one of the cells firing B, is increased.

(Hebb, 2002)

Hebb's concept is often paraphrased to 'Neurons that fire together, wire together'. In fact, it was a Spanish neuroanatomist in 1894, Santiago Ramon y Cajal, who first proposed a way that we could learn that wouldn't involve the creation of new neurons. The exciting thing is that we have the power to influence the very structure of our brains. For coaches this is key. This could well be the most important chapter in this book, and content that should be included in any book on coaching! Certainly every coach and manager should have a thorough understanding of this topic.

Experience produces changes in the brain, some of the changes that occur being:

- increased dendritic length;
- increased dendritic spine density (or decreased);
- synapse formation;
- increased glial activity (commonly considered as the glue of the nervous system);
- altered metabolic activity.

There is not scope here to go into all the research that suggests the variety of options open to coaches, leaders and managers to help encourage neuroplasticity. However, we will touch back on a selection of strategies in later chapters.

Why is it important to me as a coach/ manager?

Neuroplasticity takes the centre stage here in the relatively recently recognized ability of the brain to change. From the micro changes in neural pathways right up to the macro effects typically accompanying a brain injury we are now convinced that the brain can physically change. This fundamental process is at the heart of most of the work that coaches do with their clients. In fact, we believe that coaching facilitates 'self-directed neuroplasticity'. When we share that belief during the Neuroscience for Coaches training programme most coaches get quite excited because they see the power of neuroplasticity. It underpins the basis of change work. The goal of a coach is to help people direct plasticity to help shape their brain, or optimize their brain to enable them to fulfil their potential. In organizations this is what we spend a great chunk of our time educating teams and leaders to do.

There are many old and fascinating studies in this area conducted by people such as Paul Bach-y-Rita and Michael Merzenich. Some of the early research involved things such as helping blind people to read and perceive shadows – this was dubbed 'the first and boldest applications of neuroplasticity'. It involved an electrically stimulated chair with a large camera that sent signals of what the camera saw to stimulators against the subject's skin. The results were as profound as the individuals eventually being able to recognize a picture of Twiggy! The profoundness of these results is hard to comprehend. People who could not see were able to process visual information – the brain behaved with a huge amount of plasticity.

We explore several studies around neuroplasticity in *Make Your Brain Work* (2020), including some of the work by Michael Merzenich. He is very bold in his understanding of what neuroplasticity can accomplish. He suggests that brain exercises may be as useful as drugs to treat diseases as severe as schizophrenia.

An experiment was done that focused on the plasticity of working memory (Li *et al*, 2008). The ability to hold things in our working memory for short times is a very valuable one. The study looked at what effect extensive working memory practice had on performance improvement, transfer and short-term maintenance of practice gains and transfer effects. Two age groups were included, 20- to 30-year-olds and 70- to 80-year-olds. They all practised a spatial working memory task with two levels of processing demands for 15 minutes per day for 45 days. The results showed substantial performance gains (so they got a lot better), near transfer to a more demanding spatial and numerical task (so they were able to transfer their skills to harder tasks) and three-month maintenance of practice gains and near transfer effects (so the results lasted).

A study that as an ex-medical student I find amusing, as well as fascinating (and experientially believable), is one that focused on the brains of medical students (Draganski *et al*, 2006). These volunteers had their brains scanned before and after studying for their exams. In this short period of a few months their grey matter increased significantly in the posterior and lateral parietal cortex.

A case that may be motivating to many people is that of Rudiger Gamm, a German who became a mathematical phenomenon. Born with no special powers, he can now calculate the ninth power or fifth root of numbers and solve problems such as 68×75 in five seconds! From age 20 he spent four hours per day doing calculations, resulting in him at the age of 26 being able to earn a living performing such calculations on television. Positron emission tomography (PET) scans revealed that he was using five more brain areas than was the norm for these calculations.

Neuroplasticity underpins learning, remembering things and changing behaviours. Experiences can change both the physical structure of the brain and the functional organization of it. This in turn affects how we respond to experiences. One of the most well-known examples of this is the practice of meditation, which has been shown to increase the actual physical size of different areas of the brain as well as connectivity between regions. We will look at this more in Chapter 27 on mindfulness.

Some things you could try with your clients or your colleagues include eliciting something they would like to work on, for example being more patient with their children, then focus for two months on strengthening the neural networks associated with patience. This might include visualizing situations where previously they would have become frustrated but instead respond in a calm manner, taking a deep breath before they respond to their children, doing things that they know will require patience and talking to people about successes they've had with being patient. There is good reason to let people 'sleep on it'. When we sleep we consolidate learning and memory. Sleep is also a great time for plastic change to occur in the brain.

Opportunity

There is actually a wonderful opportunity with neuroplasticity. This is a concept we, at Synaptic Potential, almost always mention when working with organizations. Many people have heard of it now, but still not many feel confident explaining it and the implications. I hope that in the future everyone from the very young to the very old has a real understanding *and experience* of what this amazing property means for them. When people get it, it can free them, inspire them and motivate them. Don't just scratch the surface of neuroplasticity. Consider the deep opportunities.

Helpful insights for managers who coach

The mindset that accompanies the scientific fact of neuroplasticity is the growth mindset – researched and popularized by Carol Dweck (2007). Many organizations are keen to embed and value a growth mindset. This is great, and a good start... but saying it is an organizational value or even priority alone won't unlock the potential of people. The habits, the behaviours, the structures and organizational policies and procedures all must be aligned. The organization has to ooze growth mindset.

Threat/reward response   18

What is it?

The threat response is a series of reactions that occur within the body when we *perceive* a threat – the same as if there were a real threat present. We use the term 'threat' to include a range of subtle triggers, which have not all been conclusively researched. When we work with groups of leaders there can be confusion between this and the fight-or-flight response because the basic physiological response is similar; however, the physical effects of the threat response are normally less stark, while the mental ones remain potent.

Why is it important to me as a coach/ manager?

The threat response underpins so much of a client's day, unconsciously, that it can drive a lot of their actions. It is the antithesis to a High-Performing Neural Environment. Understanding what some of the triggers are and how they work enables you to ask better-quality questions and direct attention more effectively. There are many studies that look at the effects of various triggers of the threat response that we cover in various other sections of this book. By making your client and colleagues aware of potential triggers, and how triggers may compromise others, you are enabling them to be more effective in a huge range of situations.

For managers in a coaching role it is especially important to realize that we know that when the threat response is activated the body is flooded with norepinephrine, which can be useful for increasing focus. However, the whole stress response overall isn't great because it:

- inhibits creative thinking;
- tires the brain quickly, reducing cognitive resources;
- can have negative health impacts due to increased cortisol;

- impacts on your ability to think;
- can make you behave or think more defensively;
- can make you less able to process new information.

A stress response that comes and goes isn't problematic. The threat response is very powerful and can be long lasting. Much more so than the 'toward' response you get with a potential reward, which is sometimes disappointingly short.

Ultimately these responses are hugely important because they have a direct effect on the efficiency, effectiveness and productivity of people. For example, if an employee feels they are being micromanaged by their manager it can directly affect their productivity, which ultimately costs the company money. So coaching leaders to effectively work with people and avoid triggering the threat response is vital.

The neuroscience of the threat response

Pressure in itself isn't bad. (A common challenge managers put to us – arguing that they work really well 'under pressure'). But a person with the neurological responses of fear, threat and a lack of safety or security isn't going to be able to perform to their best. The threat response can be thought of as a collective term for the bringing together of lots of research that identifies a below optimal performance from individuals. As we have noted, some of the physiological responses are similar to those of the fight-or-flight response.

The neuroscience of the fight-or-flight response is well documented. The autonomic nervous system is made up of both the sympathetic and parasympathetic systems. It is regulating many unconscious processes every minute of every day. For example, it controls our heart rate, digestion and breathing rate. The sympathetic nervous system activates the release of norepinephrine when a threat is presented. The parasympathetic system can be thought of as the restorative system releasing acetylcholine and bringing back a homeostatic balance.

In the fight-or-flight response the amygdala triggers a neural response in the hypothalamus. The pituitary gland is then activated and secretes adrenocorticotropic hormone. Around the same time, the adrenal gland is activated and releases epinephrine. Cortisol is produced and this has various effects, including suppressing the immune system, increasing blood pressure and increasing blood sugar levels. This creates an energy boost, along with an increase in heart rate and breathing rate, preparing the body for fight or flight.

Turning our attention now to the threat response. When a person experiences a threat response a series of events inside their body are triggered. The process itself uses up oxygen and glucose. This has the effect of decreasing working memory capacity. You need your working memory for creative insight, analytic thinking, problem solving and even just simply holding something in your very short-term memory. These are all impaired when you have a threat response.

A threat response also affects the amygdala, anterior cingulate cortex and frontal lobe. The release of cortisol is triggered and this results in decreasing immunity, impairing learning and affecting memory. Together, this spells disaster for efficiency, effectiveness and productivity.

The brain cannot distinguish between real and unreal threats. For example, if you see what may be a snake near you on a country walk it is safer to respond as if it is one, rather than presume it isn't. Your midbrain region responds quickly to keep you safe. If it turns out that there isn't a snake, your brain usually calms down the threat response. In organizations things are normally far more complicated and your brain often struggles to determine if something is a real threat, so it keeps you safe by presuming it is!

When a threat is perceived, just like in the fight-or-flight response, the important system called the hypothalamic-pituitary-adrenal (HPA) axis kicks into action. This system comprises interactions between the three named brain areas. The HPA system triggers the production and release of cortisol and epinephrine. This prepares the body for action; the blood pressure and heart rate increase, the lungs take in more oxygen by increasing the breathing rate and blood flow can dramatically increase.

Neurotransmitters are also released as a result of the HPA system's activation. These chemical messengers activate the amygdala, which then triggers the brain's response to emotions. The chemical messengers also tell the hippocampus to create a record of this emotional experience in the long-term memory. Finally, the neurotransmitters also suppress activity in the frontal lobe, which means that short-term memory, concentration, inhibition and rational thinking all take a nosedive. Trying to handle social interactions elegantly or do cognitive tasks becomes very challenging while in this state.

Suggestions

The most important thing with the threat response is to educate people that it exists. Once people are aware that it can be hampering their efforts in many different areas of their life they are then in a position to do something about it. In the workplace imagine the difference if all leaders, managers and team members knew that this was a possibility for people they were dealing with. The opportunities for increased empathy, patience and understanding would grow.

Potential triggers of a threat response include:

- a perceived lack of control;
- feelings of uncertainty;
- any sense of unfairness;
- poor connectedness.

So both exploring where these may be occurring in your client's or colleague's life and also what may be going on for other people in their lives at certain times – friends, other colleagues and family – could be useful. It could also be worth identifying ways to strengthen any recurring areas of susceptibility. For example, if someone often experienced a threat response at work due to a perceived lack of control there might be steps that could be taken to reduce the frequency of this. Is it because their boss has a tendency to micromanage them? (If so, could they have a conversation with the boss?) Is it because customers keep changing their mind about what they want? (If so, could procedures be put in place to minimize this?) Or is it because they are filtering for the things that they cannot control? (If so, would some work on focus and filters help?)

Helpful insights for managers who coach

Protecting your team members' ability to think, process, be creative and be productive is a main priority for you. It often isn't in job descriptions though. 'Creating psychologically safe environments' is yet to become a mainstream skill that managers are asked to demonstrate. However, it should be. Your people can only reach their potential, and only do their job at a high level, if you are helping create these High-Performing Neural Environments rather than hindering them.

Neuroimaging　　19

What is it?

The field of neuroimaging is a specialized one. It includes all the techniques currently at our disposal to study the structure and functions of the brain. These two main categories, structure and function, enable us to focus in on different things. Structural imaging looks at brain anatomy and is used in a medical context for diagnosing things such as lesions or disease. Functional imaging looks at brain activity and so it is used for the cognitive psychology research that you will read a lot about in this book. We talk about brain areas 'lighting up', this is because there is an increase in metabolism in a specific area and it then shows up with definition on functional imaging scans – we then know that this area of the brain is processing information.

So what are some of the research tools that neuroscientists use?

By purpose of a historical overview we'll start with the computerized axial tomography (CAT or CT scanners). These were introduced in the 1970s and gave us our first, now familiar, pictures of the inside of the body, which were useful diagnostically and for research intent. The inventors, Allan Cormack and Godfrey Hounsfield, received the 1979 Nobel Prize for Physiology or Medicine for this and it was a real boost to neuroscience. Shortly afterwards, the single photon emission computed tomography (SPECT) scan and positron emission tomography (PET) scan were facilitated by the development of radioligands. This was in the early 1980s, when the magnetic resonance imaging scan was also being developed. Its use initially was mainly clinical. Functional magnetic resonance imaging followed shortly after this and, in 2003, Peter Mansfield and Paul Lauterbur were awarded the Nobel Prize for Physiology or Medicine for the development of the MRI scanner.

Computerized axial tomography (CAT or CT)

This form of scanning uses X-rays. We see an end result of pictures that look like many cross-sections of the brain. The bit in between involves a computer program that performs a numerical integral calculation estimating how much X-ray beam has been absorbed. The denser a part of the brain is, the whiter it will appear. This is just like 'normal' X-rays where bones show up as a strong white and we can then easily (well, easy-ish) see where there are fractures because they show up as dark lines. CT scans have most value when a picture of the brain is needed quickly because there are concerns about swelling or bleeding. You can get good images quickly with a CT scan, whereas with fMRI the process takes a lot longer. CT is accurate, reliable, safe and so really good for people who have experienced a head trauma.

Positron emission tomography (PET)

This involves the use of a radioactive chemical (tracer) that collects in cells that are using a lot of energy and a special camera that can record the positively charged particles that the tracer gives off. Individuals are injected with this compound and it travels through their bloodstream up to the brain. The resulting scan is a multicoloured two- or three-dimensional image that highlights where the compound acts in the brain. Different compounds can show blood flow and oxygen and glucose metabolism in the tissues. The scans are quick, indeed they can only measure short tasks as the radioactivity decays quickly.

The picture isn't as detailed as MRI or CT scans and tends to have clinical uses, such as looking for cancer, Alzheimer's disease and strokes. These all cause observable changes in metabolism so are easily picked up by PET scans. PET scans are able to identify specific receptors associated with particular neurotransmitters. This is a unique and valuable feature.

Single photon emission computed tomography (SPECT)

This also involves administering a compound to the person, in this case a gamma-emitting radioisotope. We also get a two- or three-dimensional image of the brain regions that are active. This type of scan is very useful for people with epilepsy, because once you inject the radioactive tracer it is distributed within the brain 30–60 seconds later. You can then see what the cerebral blood flow was when the tracer was injected. The downside is that the scan has poor resolution.

Magnetoencephalography (MEG)

This measures the magnetic fields produced by electrical activity in the brain. MEG delivers a direct measurement of neural electrical activity. The temporal resolution is high and the spatial resolution is low. The results from MEG are considered less distorted than from electroencephalography (EEG). MEG is used for research, in surgery to localize pathology and for neurofeedback, among other uses.

Functional magnetic resonance imaging (fMRI)

Functional magnetic resonance imaging uses magnetic fields. Due to the use of magnetic fields there are some people who cannot go into these scanners. People with cochlear implants, pacemakers or other surgical implants, some forms of tattoos and people at risk of having shrapnel in them.

It measures the proportion of deoxyhaemoglobin in the blood, which is correlated with brain activity. The paramagnetic properties of oxygenated and deoxygenated haemoglobin are key to how we are able to see the change in activity in the different areas of the brain. Resolution is 2–3 millimetres, which is great. People can be asked to do simple tasks while in fMRI scanners, for example pressing a button; this has dramatically increased what we can do from a research perspective. For coaches, the fMRI scanner is used most in the types of studies we are interested in. Most of the studies you will come across (certainly the ones we tend to refer to) utilize fMRI scanners.

Why is it important to me as a coach/ manager?

The difference between a coach who has a depth of understanding versus one who has only a surface-level understanding is significant. There are many benefits to really getting to grips with this information, and a foundational piece of knowledge is about knowing how data is gathered. It is similar to going into an organization and being given a report about someone. You would rightly want to understand how that information had come to be on the report. Did the individual self-report after some reflection? Was the report the result of a 360-degree evaluative process? Was the report written by the individual's boss? Or was it compiled from an exit interview from a disgruntled previous team member? As you can see, where the information came from can dramatically affect the filters with which you read the report.

If you choose to give yourself a thorough grounding in the foundations of how the brain works by attending a programme on Neuroscience for Coaches then (at least from our programme) you will be equipping yourself to read scientific papers in the future. Knowing how the equipment that helps gather the data actually works gives you an insight into the scope of interpretation from that data.

Working memory

What is it?

'Working memory' is a term that psychologists came up with to describe our ability to hold things in the front of our mind and manipulate these small chunks of information. A classic example is mental arithmetic. The term is often used synonymously with short-term memory, and is thought by many to supersede the older 'short-term memory' term. When a task is performed using the working memory system it involves executive and attention control enabling integration, processing, disposal and retrieval of information. It is important to remember that this is a concept that at this point is theoretical, a working model that is useful to our understanding of things.

We believe that frontal cortex (especially the dorsolateral prefrontal cortex), parietal cortex, anterior cingulate and parts of the basal ganglia are important in short-term or working memory. The term 'working memory' was originally coined in 1960 by George Miller, Eugene Galanter and Karl Pribram and was linked to the very computer-like model of the mind. When we talk of using our working memory we are talking about a couple of seconds; it has a limited capacity. In 1956 Miller proposed that we could hold seven units in our mind. In neurolinguistic programming (NLP) this is referenced as the 'magic 7 plus or minus 2' chunks (Miller, 1956). We know now that there are several factors that affect how many chunks a person can hold in their working memory. For words, for example, this is dependent on the length of the words and the familiarity of the words and the memory strategies used.

Why is it important to me as a coach/ manager?

People use their working memory very frequently. There are ways to engage with it to help optimize its performance, and things that are unhelpful when working with it. Being aware of what it is and how it works means you can help people notice strategic changes they could make. Perhaps you might even want to invite your clients or colleagues to do certain things in order to improve their working memories.

As you have just read, the 'people can remember 7+ or −2 chunks' has since been revised. Various factors are involved in how many 'chunks' a person can hold in their working memory. Something to bear in mind when we are discussing working memory is the research of K Anders Ericsson, who has studied individuals who have been able to remember up to 80 chunks (Ericsson, 1980). This, and similar memory experiments, can be quite deceptive though as this individual had in his long-term memory lots of digits from other reference points, in this case racing times. Ultimately at the working memory level there was a relatively small number of chunks being stored, but each of these unpacked a lot more digits below. Almost like a code or a parcel that unwraps (or acts as retrieval cues) and gives a lot more data at the next level. Using memory-training skills therefore would not increase working memory. Indeed this individual with a great digit memory had normal word-recall ability.

A randomized controlled study showed that working memory training increased a range of cognitive abilities and that brain activity in the prefrontal cortex increased. Also, measurable increases in the density of dopamine neuroreceptors were noted. From the neurotransmitter perspective both dopamine and glutamate are thought to be involved in working memory.

A key study, for example, highlighted how important our ability to suppress irrelevant information is for our working memory to work optimally. Distractions need suppression, which often means that the neural activity in the sensory cortex needs modulating. In very simple terms, our ability to remember relevant information depends on our ability to limit interference from irrelevant information. This hopefully sounds very logical and resonates with you. Until studies are done to confirm the mechanisms that are occurring at a neural level we don't know some of these things for sure, we are only guessing that what sounds logical on an experiential level could be mimicking what is occurring at a brain cell level. It has now been shown that when we are distracted by irrelevant information the neural process involves an increased memory load. Our ability to effectively filter irrelevant information by neural suppression, in order to prevent overloading our limited working memory capacity, is a key process.

Coaching provides a wonderful opportunity for self-evaluation. Inviting a client to give themselves an honest appraisal of where they feel their current ability to suppress irrelevant information lies could be a very valuable experience. Considering different areas of life – for example, at work when on our own, at work when with others, at home on our own, at home with partner, with children, etc – could bring to light some opportunities for development. Of course there are environmental things we can do to help reduce the need for the suppression of external distractions. One of the easiest at work or home would be to tidy up. Clear out all old stuff that you are not currently using. Put things out of sight that are not in regular use.

Managers have a big responsibility to create environments that are as mentally tidy as possible. Overloading people with tasks, giving them information they don't need, inviting people to meetings 'just for information' are all bad news when it comes to optimal cognitive processing. If you want people to give you all the benefits of a High-Performing Neural Environment, don't clog up people's working memory with unnecessary data.

So we know that our ability to store things in our working memory and process information there is impaired when we allow distractions to do their job. The question could then become: what can we do about that? The answer lies partly in a study that looked at practice-related improvements in working memory. This study used electroencephalographic (EEG) recordings with individuals while they performed three motion-direction, delayed-recognition tasks. The first task didn't have any interference associated with it, the other two involved different types of interference during the interval of memory maintenance: distractors and interruptors. Both of these forms of interference disrupted working memory performance. However, both accuracy and response time improved with practice. Thus it would appear that we could get better at filtering irrelevant information.

One favourite study looked at the effect of chewing on working memory (Hirano *et al*, 2008). Chewing has been linked to the activation of various brain areas, including the prefrontal cortex. This study looked into the specific effects of chewing on neuronal activities during a working memory task and used fMRI to help see what was happening. The participants of the study had to chew odourless and tasteless chewing gum and perform specific tasks of working memory. The results were complex, showing activation in lots of areas of the brain, including the dorsolateral prefrontal cortex. There was also the suggestion that chewing may accelerate the process of working memory.

For anyone working in a knowledge-based role, where their ability to think or to compute are important assets to them, there is room to support growth. Consider teaching your client or your colleagues what working memory is and what its limitations are. Remember:

- When we get distracted we often lose the contents of our working memory.

- If someone is distracted when giving us important information then the information is unlikely to even get into the working memory – so it hasn't even been properly heard... let alone registered or remembered.

- Your working memory has a limited capacity – so expecting it to hold lots of information is futile. Also trying to do complex things, such as 279×534, for most of us would be tricky because of the number of steps and different chunks of information that need to be stored.

- Everyone is different – sometimes this may be down to differing working memory capacities, but sometimes, like in chess or mental arithmetic, it is also down to a different type of memory being utilized.

- High-intensity exercise has been shown to increase the capacity of working memory.

- Stress impairs working memory – so if people are stressed they can't expect to have their working memory in optimal working condition.

Overall, it is suggested that by improving our working memory we can become happier, more focused and more effective. Some odd ways you could improve it include:

- Climbing a tree – you'll use your working memory to balance and plan your route while evaluating your current position.

- Eat some dark chocolate (temporary boost).

- (The big one) turn off your smartphone – the constant interruptions are deadly to prolonged focus and getting the most out of your working memory.

Multitasking has been correlated to the processing and storage components of working memory capacity. The more effective a person's working memory is, the better at multitasking they appear to be. This is especially useful in certain job roles. We know that practice can have a big effect on the brain, so investing time in becoming really good at focusing on one task at a time could increase the synaptic connections in areas of the brain involved in working memory.

HPA axis 21

What is it?

The hypothalamic-pituitary-adrenal axis is part of the neuroendocrine system that is involved in our reactions to stress. It is also important in regulating many other processes such as digestion, the immune system, mood and emotions, energy storage and expenditure. Essentially it is made up of a series of detailed interactions between the hypothalamus, the pituitary gland and the adrenal glands.

The hypothalamus is responsible for making and releasing vasopressin and corticotropin-releasing hormone (CRH). Together they regulate the pituitary gland. Adrenocorticotropic hormone is released, which acts on the adrenal cortex. This produces cortisol, among other things. They then act on the hypothalamus and pituitary in a negative feedback cycle.

CRH is released from the hypothalamus in response to several things including stress, physical activity and illness. It is also linked to the circadian rhythm of an individual. You'll remember from Chapter 10 on cortisol that levels rise rapidly after we wake and reach a peak after 30–45 minutes. Abnormalities in this cycle are linked to insomnia and burnout.

Connections between the amygdala and the hypothalamus mean that sensory information that arrives at the amygdala can be communicated to the hypothalamus quickly. So if we see, hear or in any way perceive something that is interpreted to generate a fear response then the HPA can be activated, along with the sympathetic nervous system. As we've discussed previously, in small doses the release of glycocorticoids (of which cortisol is one) is important, but released in excess damage can be done. The hippocampus can be affected, so memory can be impacted.

Why is it important to me as a coach/manager?

HPA is involved in many disorders, including anxiety disorder, burnout, irritable bowel syndrome, alcoholism and insomnia (also many others but it is unlikely that you would be working with people suffering from many of them). The overactivity of the HPA is also said to be a characteristic of depression. Antidepressants act to regulate the HPA axis function.

Neurotransmitters such as dopamine, serotonin and norepinephrine have a role to play in the regulation of the HPA axis. Studies show that positive social interactions that increase levels of oxytocin suppress the HPA axis and this can have benefits that include improved wound healing. Being aware of the positive and protective impact that oxytocin can have is important when working with individuals who are at risk of chronic stress.

One study addressed the question 'Do burned out and work-engaged employees differ in the functioning of the hypothalamic-pituitary-adrenal axis?' (Langelaan *et al*, 2006). The saliva of just under 100 managers was collected on several days and the levels of cortisol and a compound that counterbalanced cortisol were measured. The group that were work-engaged were found to have stronger cortisol suppression in response to the other compound, which suggested that they have higher feedback sensitivity. So work-engaged employees do differ slightly in the functioning of the HPA axis. You can read more about creating workplaces where employees are engaged in 'Engaged, the neuroscience behind creating productive people in successful organizations'.

Mirror neurons   22

What are they?

The mirror neuron is a type of neuron that fires when an action is observed in another neuron. The story of how mirror neurons were initially discovered is an interesting one (*New York Times*, 2006). In the 1990s in a lab in Parma, Italy a group of neuroscientists were looking at the neurons of monkeys. The monkeys were wired up to a machine that made a noise when they reached for/ate a peanut. One day, after a graduate student came back from lunch (still eating his ice cream) the machine made the noise... but the monkey hadn't moved. This was strange because the expectation was that the neurons only fired and triggered the machine when the monkey moved to eat something. What they found was that the neuron fired when it observed others eating, as well as when the monkey ate. We now know that this occurs in humans too. Initially when the researchers tried to report on their observations their first paper was rejected. This is one of the great, and challenging, things about scientific fields such as neuroscience. It is often only with the building of information that new ideas become accepted. Currently there is still speculation around this system and it can be useful to think of mirror neurons as part of the action-observation network.

The action-observation network (AON) has been suggested to support our understanding of others' actions and goals. People can track and in part replicate the mental and bodily states of others. Neuroimaging studies have shown overlapping activation in the brain for perception and execution of motor actions in a parietofrontal network involving several brain areas. The AON is suggested to support social perception by interacting with the somatosensory cortices.

A study led by Lauri Nummenmaa published in 2014 (Nummenmaa *et al*, 2014) involved subjects watching movies showing boxing matches while in an MRI scanner. They either watched the movie passively or simulated the feelings of one of the boxers. When simulating the feelings, multiple brain regions were activated and the interconnectivity of these regions strengthened during the process. The researchers proposed that sharing a third person's feelings synchronizes the observer's own brain mechanism's supporting sensations and motor planning. This makes mutual understanding more likely.

It is suggested that similar brain states could be a prerequisite for similar mind states. This makes logical sense. Other studies have shown overlapping neural activation in perception, experience of emotional states such as pain, disgust and pleasure.

For many years it was thought that mirror neurons were only related to motor functions. Now it is thought that, in Dr Rizzolatti's words: 'Mirror neurons allow us to grasp the minds of others not through conceptual reasoning but through direct simulation. By feeling, not by thinking' (*New York Times*, 2006). These special mirror neurons have been found in the premotor cortex, posterior parietal lobe, superior temporal sulcus and insula. When we are in a meeting and someone is struggling, either forgetting things or having to share bad news, people often 'tune in' to what is going on for that individual. They feel what that person is feeling. Some neuroscientists say that this is down to mirror neurons.

In recent years some neuroscientists have gone a step further than thinking that mirror neurons only have a role in mirroring motor actions, and have suggested that mirror neurons are also important in:

- theory of mind skills;
- helping us to understand the actions and intentions of others;
- emotions such as empathy.

These obviously are involved in a lot of our daily lives and so awareness of a potential influencer is important. If mirror neurons do help us to understand the intentions of other people then this is another reason to support people in trusting their instincts. Strengthening trust in oneself is also important if mirror neurons can pick up on and help us to perceive the feelings of another person and so enable us to be more empathetic. Right now, however, the official party line (if one exists) is that there is not a widely accepted neural model to explain how these things may work. Yet that does not mean that in time there won't be such a model. But to ensure you know the flip side of interpretation out there consider the work of Gregory Hickok who considers mirror neurons a myth.

A study by Christian Keysers (2011) found that when people watch a hand extend to caress someone but be intercepted by another hand pushing it away (rudely) then the insula of the observer registered the social pain of rejection. It should be noted that the existence/extent of mirror neurons outside of motor neurons is strongly debated. Studies are cropping up every year and the neuroscientists are continuing to thrash out their differences as you read this!

Giacomo Rizzolatti, Christian Keysers and others have been investigating the role of the anterior insula and anterior cingulate cortex in order to better comprehend how we understand the emotions of others. In one study the researchers looked at the fMRI scans of participants while they inhaled odorants that produced feelings of disgust in them (Keysers *et al*, 2003). They also then watched video clips of people showing the facial expression of disgust. The results showed that both seeing disgust in others and feeling disgust oneself activates the same sites in the anterior insula and to a lesser degree the anterior cingulate cortex. Keysers says that there is a uniquely human mirror-neuron system found in the insula. He also says that people who are highly empathetic (as ranked on a scale measuring empathy) have particularly active mirror neuron systems.

Why are they important to me as a coach/ manager?

Consider this example: there has been a stressful meeting and everyone is tense. Bad news has been delivered to the people around the table. Before anyone realizes it, people have started to attack each other. Your client focuses on exploring the surface-level challenges, thinking Jo is wrong or Bob needs to behave differently. The surface-level challenges are likely not the issue, however. It is possible that mirror neurons are creating an intense group of stressed individuals, the core cause being fear rather than any surface problem. As one person responds in an aggravated fashion, others do likewise. This isn't necessarily down to mirror neurons at all, although it just might be possible. We do know for sure that emotions are contagious. As a manager one of your opportunities is to be the emotional tone setter. What emotions are most helpful to create High-Performing Neural Environments for what your team needs to accomplish?

When you are working with a client could you conduct a session focusing on the external influences that may be affecting them? Do they read the newspaper each morning, filling their mind with pictures of disturbing things that may be putting them in a less than optimal working state... is there a better time to read that newspaper? We know that repeatedly activating areas of the brain strengthens neural connections and can increase the size of different brain regions... so what are they frequently looking at? For managers, how are you setting your team up for the day? What is the energy like in your meetings?

PART FOUR
Brain networks

Many of us like things to be simple. We like to be able to break things down into chunks that are easy to understand. We have started with that approach to the brain, but it isn't the whole story. While the brain has fixed anatomy (as opposed to the heart with moving parts) it has many functions (the heart has one function). The brain is very complex. This complexity of connectivity of the network architecture enables us to respond to the range of environmental demands we are subjected to.

The networks in the brain are so important to our understanding of how we really work. While the realities of the complexities would take a fair time investment to get to grips with, we have created a simplified model called 'The Home Grown Brain'. Do reach out to us at Synaptic Potential if your organization wants to learn more about how understanding brain networks can help teams achieve High-Performing Neural Environments.

Brain networks   23

Looking at the brain from the position of its networks is a different approach and one that offers another level of understanding. Our comprehension is expanding in neuroscience at a rapid rate. Concepts that were once favoured – some for their ease of understanding and applying, for example, the idea of people being 'left or right brained' – perhaps didn't have solid research underpinning them and are falling away. We are left with the myths, though.

Hot and cool

In psychology a dual system is quite widely explored, with the foundations of the thinking being attributed to William James. Many theories have followed, among them Daniel Kahneman's interpretation in 2003 of this hot and cool 'system 1' and 'system 2' way of processing information.

System 1 is implicit (automatic) and unconscious, whereas system 2 is explicit (controlled) and conscious. You are probably familiar with the classic experiment where children are asked to sit with a treat such as a marshmallow in front of them and decide whether to eat it now or wait for 15 minutes to receive another one. The researchers suggested that this was an example of the conflict between the two systems and that most children are challenged because their cool system is underdeveloped.

From a neuroscientific perspective we can use MRI to look for the possible underpinnings of these networks in order to add another layer to our understanding. In a study led by McClure (McClure *et al*, 2004) functional MRI was used to observe brain region activation while the participants in the study made decisions about monetary rewards. The results showed that when choices involved immediate gratification, areas of the midbrain were activated. Regions involved in emotion, behaviour, motivation and long-term memory. When participants were making choices that involved delaying gratification, the cortex was more active. This is the evolutionary newer area of the brain that forms the outer layer of its anatomy. The regions here are important for cognition and regulation.

We can think of system 1 (or the 'hot' system) as being emotional, unconscious, automatic (so fast), simple, evolutionarily old and nonverbal. Whereas system 2 (or the 'cool' system) is cognitive, complex, self-controlled, slower, involves effort, logical and is limited by our working memory capacity.

This network is especially important when we are considering:

- Choice architecture – which can utilize the hot network to get immediate results.
- Habits – where we are often operating in system 1.
- Decision making – where system 1 could be playing a bigger part than we realize.
- Expectations – the dance that occurs beneath system 2's radar.

Really understanding this network and how it could be influencing people you're working with is vital. You won't have a High-Performing Neural Environment if your people's behaviours are being dictated by a poorly programmed hot network. It takes time to change things, but the first step is awareness. In many of our programmes with organizations we are supporting the raising of awareness and the utilization of the hot network for desired outcomes.

Default mode

The default mode network is the collection of the areas of the brain that increase in activity when we are not focused on a task. It is sometimes referred to as a resting state network. The areas include the medial prefrontal cortex, medial parietal cortex, posterior cingulate cortex and medial temporal lobes. The research is in its early stages and so some of the consequences of these cortical associations have yet to be uncovered.

When we have thoughts that are independent from our immediate environment the default mode network appears to be active. Examples include when our mind is wandering, when we are being introspective or when we plan what we're doing some time in the future.

One of the theories about default network activity is that it is the brain's baseline of processing and information maintenance. The suggestion is that the network helps us consolidate our experiences and prepare to effectively react. Interesting people with specific medical problems such as attention deficit hyperactivity disorder (ADHD), autism, Alzheimer's, schizophrenia and other mental illnesses have been seen to have different activity in this network. Additionally, people who have experienced long-term trauma, such as abuse, appear to have lower connectivity across the default network.

Dorsal/ventral attention stream

The final network we will look at here is a model for the neural processing of vision which is linked to our ability to pay attention (vital for work!). As with many models it is a work in progress; the components are useful to us now, and further research may develop the model. It suggests that after visual information has gone into the occipital lobe (the visual-processing lobe of the brain, found at the back of the brain and home to most of the visual cortex) the information then follows one of two main streams.

The ventral stream is also known as the 'what pathway'. It is involved in us being able to identify and recognize objects and subsequently plan behaviour. The other stream is known as the 'where pathway' or the dorsal stream. It is involved in processing the location of objects relevant to us in order to programme our behaviour. The ventral stream travels to the temporal lobe, while the dorsal stream heads to the parietal lobe (two of the four lobes that the brain is divided into, the others being the occipital and the frontal lobes).

So we have one pathway providing a detailed representation of the world for cognitive processes. It is relatively slow, involves long-term stored memories and there is typically a high level of consciousness around it. The other pathway is translating the visual information into coordinates required for skilled motor behaviour – it helps guide us. It is relatively fast, involves only very short-term memory storage and typically involves low consciousness.

These streams are a beautiful illustration of how brain areas work together. Remember that the brain is not designed as separate components. We observed it, cut it up and labelled different sections. When we see the passage of visual information from our eyes into our occipital lobe, where visual functions are happening to the spatial awareness functions in the parietal lobe (as is the case with the dorsal stream), we get a real feel for the networks that are present in the brain. Considering that the areas in the ventral stream are influenced by factors such as attention, working memory and the salience of stimuli, we can see that its role is also in evaluating the significance of what we see.

Network conclusion

The real opportunity available to us after looking at some brain networks is to be open to this different way of looking at the brain. The temptation is to focus on 'what bit of the brain does this?' when rarely is it that simple. Just like in organizations where several departments often talk to one another, work on the same project or socialize together, the brain is also very interconnected. In the Neuroscience for Coaches training programme and the Coach Savvy Manager programme we go into several other networks that have real practical relevance for how you can help people get the best out of themselves.

PART FIVE
The quantum brain

This is a very special part of the book. Often when working with organizations it is interesting to note that in all types, from large corporate ones right through to smaller entrepreneurial ones, individuals get very excited when quantum physics is mentioned. Originally this section slotted into Part Three of the book; however, after reflection it got its own section. There are a couple of reasons for this. One, the content of this section is quite controversial. Speak to many physicists and they will tell you that several things mentioned are not possible. Speak to many neuroscientists and they will tell you that these lines of thought are not going anywhere. However, if there is one thing that science has taught us down the ages, it is that we can be wrong. So in the next 10–25 years one of two things is likely to happen: either this section will be removed from future editions due to additional research rendering it unworthy, or it will link to a whole new book, which expands what we currently think. Personally, I believe it will be the latter, which is why it is included here.

The second reason that the quantum brain got its own section is that there is a lot covered here and before reading it you will want to ensure you have the head space to dive into all the deep content. So take the opportunity to refresh yourself and get ready for some information that could change everything.

The quantum brain

24

What is the quantum brain?

The prevailing understanding of how the brain works, indeed how most of the world works, is in alignment with classical physics. However, there are some people who believe that what we know about more contemporary physics, namely quantum physics, is also important in our understanding of the world. For most people this is an alien concept and the implications seem quite 'out there' or unbelievable. A traditional and pure neuroscience approach tells us that the brain is made up of material particles and fields, and that everything can be explained in these terms.

Quantum physics challenges these ways of thinking. It offers experiments that show that human thoughts, choices, will is important. If, as some scientists advocate, this key concept is applicable to neuroscience then it gives us an entirely fresh conceptual framework. As with many things historically in science, good ideas often take a little time to move from the sideline into the mainstream. Some of the strong drivers pushing these understandings forward are as follows. The structural features of ion channels that are fundamental to synaptic function need quantum physics to analyse them. Quantum physics is also more appropriate in neuroplastic mechanisms that involve directed attention and mental effort.

Why is it important to me as a coach/ manager?

Contemporary physics is of importance to everyone because our understanding of our world impacts our beliefs and our values. In classical physics we can reduce everything to a mechanical process. In quantum physics the causal role of human consciousness is important in our reality. Niels Bohn, Werner Heisenberg and Wolfgang Pauli had intense exploratory discussions, the result of which was valuable new ways of thinking. In these new ways of thinking our value-based intentional efforts can make a difference in our own behaviours, our social environment and our larger reality.

In classical physics you can understand a system at a single point in time by knowing the location and velocity of all the particles within that system. You also need to know any fields that are acting on them (such as gravitational and electromagnetic). Any human's consciousness is not considered important in classical physics from the perspective of being involved in any way. Classical physics is deterministic whereas quantum physics is probabilistic.

Henry Stapp, a theoretical physicist, summarizes that in psychology what people allow into their consciousness is strongly linked to what their mind intends; people experience what they are looking for. Providing the potentiality for an experience to occur, the will of the individual draws out this from the range of options. Stapp likens this to quantum theory. He and two colleagues explore this further: Schwartz, Stapp and Beauregard (2005) suggest there are five facts that mean that it is advantageous to neuroscience to use the frameworks from contemporary physics rather than classic physics:

1 They suggest that the terms 'feeling', 'knowing' and 'effort' are intrinsically mentalistic and experiential and so cannot be described exclusively in terms of their material make-up.

2 In order to explain how big systems work, the founders of contemporary physics had to recognize the importance of choices made by humans about how they will act on the behaviour of the small components (the atoms) of these big systems.

3 This new conceptual framework has choices described in psychological language.

4 This terminology is very important to design and execute the type of experiments that demonstrate self-directed neuroplasticity.

5 What was done in atomic science could be also done in neuroscience by applying the same mathematical rules.

The exciting proposition is that this new framework offers the opportunity to allow the data from self-directed neuroplasticity experiments to be understood in a coherent, scientific and useful way. More so than previous frameworks.

Hoping you are still with me, let's explore some more of this new way of looking at the world. Rather than thinking of the world as made out of lots of bits of matter, consider instead that it is made up of potentialities.

What is quantum theory?

Niels Bohr said: 'Those who are not shocked when they first come across quantum theory cannot possibly have understood it' (Bohr, 1995). As you explore this section please bear this in mind. An important point is that contemporary physics, quantum physics, does not replace classical Newtonian physics, it adds another layer of understanding. It came about because the Newtonian principles were not working when physicists were exploring the subatomic world. The majority of physicists would say that quantum physics *only* applies to the microscopic. However, as we have seen, there are some scientists who believe that within time we will understand the macroscopic world through this lens also.

There are some occurrences at the quantum level that may challenge your current understanding of reality. If you took any physics at school it was likely to be classical physics and you were probably told that atoms – the small building blocks of matter – are mostly empty space. The analogy here is that if a basketball is the nucleus of a hydrogen atom then the electron that representationally circles around it would be 20 miles away. Everything in between is empty. Now, however, it is understood that this space is filled with powerful energy.

Another discovery that challenges widespread understandings is how electrons move from orbit to orbit. They jump. They disappear from one place in space and appear in another; a quantum jump. Even more interestingly, we cannot predict where the electron will reappear or when they will make the jump. We can only formulate the probability of these things.

Here are some key things to be aware of from the world of quantum theory:

- the double slit experiment;
- Heisenberg's uncertainty principle;
- the Einstein, Podolsky and Rosen (EPR) paradox;
- the Copenhagen interpretation.

The double slit experiment

This experiment is a fairly simple one in construct, but mind boggling as we try to grasp it. The experiment involved sending a beam of light through a thin plate that has either one or two slits in it. The important things to know are that:

- Particles can be considered to have mass, we can represent them in our mind as tiny balls. If we were to throw a tiny ball, covered in paint, at a wall we would see a tiny circular imprint on the wall.
  - Fundamental 1: when particles in the experiment pass through one slit they make a pattern on the wall behind that is in a line.
  - Fundamental 2: when particles pass through two slits they make a pattern that is in two lines.
- Waves do not have mass. Instead we think of them as oscillations or vibrations. It is useful to imagine dropping a coin into a water fountain and observing the ripples of waves emanate out from that central point.
  - Fundamental 3: when waves pass through one slit they make a pattern that is in a line (similar to the particle but not identical).
  - Fundamental 4: when waves pass through two slits they make a very different pattern; multiple lines (this is called an interference pattern).

It was believed that light was made up of particles only, discrete balls of matter. The physics-shaking shock occurred when a particle was passed through one slit and it made the pattern in a line (expected) but then went through two slits and it made the multiple line patterns (like in fundamental 4, which was very unexpected). This meant that the light was behaving like a wave, which was not at all what Isaac Newton had led us to believe.

In 1961 this experiment was done for the first time with something other than light: electrons. Our understanding of electrons is that they are particles, so they should behave in the way that fundamentals 1 and 2 dictate. They didn't. In fact, even relatively big molecules would behave as if they were waves. The magnitude of weirdness is huge at this point. The physicists tried lots of things to work it out but they just kept discovering weirder things. The particle would act as if sometimes it went through just one slit, or just the other, or neither, or both! This is like a single paint ball being fired towards two gaps and splitting into two to pass through both gaps and then rejoining again afterwards. It didn't make any sense.

Not to be outdone by a particle, though, the physicists decided to put a measuring device by the slit in order to really see what was happening. You'll never guess what happened (seriously, it's that amazing): the particle started behaving like a particle again (like fundamental 2). It is said that the observer collapses the wave function simply by observing.

Heisenberg's uncertainty principle

This basically states that we cannot know both the location and what speed a particle has: by measuring one quality we affect another.

The Einstein, Podolsky and Rosen (EPR) paradox

The EPR paradox is fascinating. A little background is that Einstein was really not a fan of quantum physics. He once claimed: 'God does not play dice with the universe.' To which Niels Bohr responded: 'Stop telling God what to do!' Einstein, Podolsky and Rosen set out to design a thought experiment that would discredit quantum physics. The idea was that you arrange for two particles to be created at the same time, meaning they are entangled (or in superposition). You send one particle off to one side of the universe and the other to the opposite side. Then you do something to one of the particles to change its state. The other particle immediately also changes its state. This was the thought experiment that they thought was pretty crazy. Remember that Einstein in his theory of relativity stated that nothing could travel faster than light? Well this, they thought, would go against that and involve communication across a universe... in Einstein's words 'spooky action at a distance'.

John Bell came up with a theory (Bell's theorem), which suggested that the EPR paradox is exactly what happens. He suggested that our concept of things being in one place, localization, is incorrect. That in fact everything is non-local and particles are linked on some level. This level is beyond time and space. On a quantum level things are connected, entangled. Schrödinger said that entanglement is *the* interesting aspect of quantum physics. Our friend Henry Stapp called Bell's theorem 'the most profound discovery of science'.

The Copenhagen interpretation

Niels Bohr lived in Copenhagen and, along with his colleagues, pushed forward the idea that the observer has an inescapable influence on physical processes. This challenging idea that we are not neutral bystanders remains a powerful one. Bohr questioned Heisenberg's assertion that things existed before they were observed. Bohr believed that particles do not exist until we observe them. Reality is created on a quantum level by observing and measuring. Einstein could not accept this anti-common-sense way of thinking. We do have a responsibility to question how common sense became so.

Lynne McTaggart suggests: 'Reality is unset Jell-O. There's a big indeterminate sludge out there that's our potential life. And we, by our very act of involvement, our act of noticing, our observation, we get that Jell-O to set. So we're intrinsic to the whole process of reality. Our involvement creates that reality' (McTaggart, 2012).

Nerve terminals

Most neuroscientists work from the assumption that classical physics will answer the questions around how consciousness and brain processes relate to one another. In order for this to be the case, classical physics has to be applicable at the microscopic level. According to Heisenberg's uncertainty principle it would be impossible, according to the principle, to gather empirical evidence. The next step is to consider the questions we should ask in order to move forward. For example, this one from Stapp (2011): 'How important quantitatively are the effects of the uncertainty principle at the microscopic (ionic) level; and if they are important at the microscopic level, then why can this microscopic indeterminacy never propagate up to the macro level?'

You'll remember from Chapter 16 on neurons and synapses the classic model of these. The electrical signal travels down the axon to the nerve terminal and then triggers the opening of channels so that calcium ions can flow into the terminal. Here there are vesicles, which are home to neurotransmitters. The calcium ions diffuse to special sites where they then trigger the vesicles to release the neurotransmitters into the synaptic gap. The neurotransmitters communicate to the neuron on the other side of the gap.

Sir John Eccles won a Nobel Prize for helping us to understand how neural communication takes place. In 1986 he proposed that the probability of neurotransmitter release depends on quantum mechanical processes. This means that the brain chemicals – which we are becoming familiar with (dopamine, serotonin, adrenaline) – are released according to some quantum laws. Some scientists believe that an observer, in this case your own mind, can influence these. In grossly simplified terms, this implies that the chemicals that flood and impact our body are affected by quantum processes, which are subject to the effect of the observer. You have an impact on the probability of your mind communicating things.

This next section gets a little more technical than we have explored before, but is key to the core concepts of the quantum brain. The channels that the calcium ions use to get into the nerve terminal are called ion channels. They are small, at some points only measuring a nanometre! This is on a similar scale to the slits in the double slit experiment. There is quantum uncertainty that any one calcium ion will be absorbed on the triggering site. Subsequently there is empirical uncertainty of a neurotransmitter being released that is affected by the quantum mixture of states of uncertainty of all the calcium ions.

It is suggested that this quantum opportunity (Stapp calls it quantum splitting) propagates to neuronal behaviour then to the behaviour of the whole brain. The argument being that bifurcation points arise where some quantum potentialities go one way and some go another. Stapp (2011) says: 'any claim that the large effects of the uncertainty principle at the synaptic level can never lead to quantum mixtures of macroscopically different states cannot be rationally justified'.

The idea that intention and attention have an impact in the brain are familiar in the fields of psychology, religion, philosophy and others but relatively new to neuroscience. The consideration that 'effort of attention' has an impact on brain activity and therefore our behaviour is a powerful one. Intentional inputs are important. Harold Pashler, a distinguished professor of psychology, mentions experiments and research that feeds into these discussions from a psychological perspective (Pashler, 1999). When a person is tasked with doing an IQ test and also giving a foot response to rapidly presented tones their mental age goes from adult to eight years old. We find time and time again that reducing effort of attention has consequences. Mental exertion reduces the amount of physical force that a person can apply. Pashler doesn't explain these occurrences in terms of quantum physics, but we can see the links.

In other experiments, this time led by Ochsner, participants are also being put to work (Ochsner, 2002). Here the participants are trained to cognitively re-evaluate emotional scenes that they are shown by consciously creating and holding on to an alternative fictional story of what is happening. They were initially shown either a negative or neutral photo for four seconds. During this time they generated a normal response to it. Then they were instructed to 'reappraise' or 'attend' to the negative or neutral photo also for four seconds. Afterwards they were asked whether they had reinterpreted the photo in order to reappraise it, or if they had used another strategy.

The accompanying fMRI scanning results showed that reappraisal is correlated with increased activity in the left lateral prefrontal cortex and the dorsal medial prefrontal cortex and decreased activity in the amygdala and medial orbitofrontal cortex. Stapp reports that, according to classical physics, when we do something like the above experiment and exert conscious effort the feeling that is linked to the process is an epiphenomenal side-effect. This has no effect on the brain. The idea that it has an effect on one's thinking is not in alignment with classical physics. The challenge is that at the micro level we know that classical physics in the brain does not fully work. In Stapp's words: 'The presumption that it gets restored at the macro level is wishful and unprovable' (Stapp, 2011).

Consciousness

Questions about the subconscious are fairly frequent when I speak to groups of people. What almost no one asks about is consciousness. Perhaps this is because people feel that they know what it is. When I was teaching in China, though, many students at a university asked me about it. They were students of computer science and know that consciousness holds the key to artificial intelligence.

There has been some movement in the field of consciousness. Sir Roger Penrose is an eminent mathematical physicist at the University of Oxford. He was awarded the Wolf Prize for physics in 1988 with Stephen Hawkins for their contribution to our understanding of the universe. Stuart Hameroff is a medical doctor of anaesthesiology. In short – two men who really know their stuff have proposed an updated theory of consciousness.

These ideas were first put forward around 20 years ago. Recently more evidence has strengthened their suggestions. To understand it we need to first meet a component of the cell structural skeleton called microtubules. They are important and complex structures. Penrose and Hameroff proposed that quantum vibrational computations in microtubules were orchestrated by their synaptic inputs and memory. Penrose believes they are terminated by 'objective reduction'.

The theory (called Orch OR) was strongly challenged by people who believed that quantum processes couldn't take place in the brain. However, there is now evidence for warm quantum coherence in plant photosynthesis, bird brain navigation, our sense of smell and brain microtubules.

These advances in research have also suggested that anaesthesia acts via microtubules in brain neurons. This is fascinating because anaesthesia selectively removes consciousness while allowing non-conscious brain activities to continue. They have said that their theory suggests 'consciousness derives from quantum vibrations in microtubules, protein polymers inside brain neurons, which both govern neuronal and synaptic function, and connect brain processes to self-organizing processes in the fine scale, "proto-conscious" quantum structure of reality' (Hameroff and Penrose 2013a and b).

Where 'the quantum brain' will be in 5, 10 and 50 years' time is largely up to the current researchers who are able to stand on the shoulders of the researchers before them to take us forward. I believe it is likely that we will continue to see more of science that demonstrates the power of our mind to affect our brain. In the meantime there are many powerful studies that add depth to what coaches are currently doing.

PART SIX
Neuroscience of classic coaching areas

Introduction

'Why write a book for managers and coaches?' I was recently asked. 'Isn't there enough to cover in a book for each audience?' There absolutely is. However, in the last few years we've been doing lots of work with managers and leaders in wonderful organizations. What I've realized through working with these dedicated people is just how critical coaching at work is. There will always be a place for external coaching. As you'll either already be aware, or soon find out, the challenge is what happens every single day. We need managers' skill sets and approaches to align more to coaches. Because the brain is being shaped daily. For me, a good manager is a good coach.

Part Six focuses on what neuroscience can share with us about classic coaching tools, techniques and concepts. These are mostly things that a coach will be familiar with and utilize in their work regularly. It can be helpful to remember that neuroscience is a field in its own right, not set up or even focused on finding things out just to back up what coaches already think and do. Therefore, sometimes we need to be careful not to be shoehorning the science into a construct that it just doesn't fit into. There are whole books dedicated to this... which you need to be wary of! Keeping an open mind and fact checking is a good way to move forward most effectively.

For both independent coaches and managers who coach we talk about the different ways people can reach their potential. One core set of pillars that need to be addressed are their people's mindframes and their skills. And a coach knows they have a whole range of tools that can be put to great use in a variety of ways. So, this section focuses on a small selection of mind-frames, skills and tools that will make the difference to anyone working on almost anything with a client or colleague.

What is coaching?

When I met up with Christian and Kim separately my first question to them both was, 'What is coaching, from your perspective?' You'll remember Christian is a professor of coaching and Kim runs a very well-regarded coach training organization. I think it is a great place to start and it's a question that even the most experienced coach doesn't answer the same from year to year – which in itself is interesting.

Christian's reply was:

> I think that's a big question and something I've been thinking about. I didn't want to pick one definition, as there are quite a lot out there. I really like Sir John Whitmore's well-known quote: 'Coaching is unlocking a person's potential to maximize their own performance. It is helping them to learn rather than teaching them.' I think, for me, it's that second sentence that I'm starting to use to explain to people, particularly in the educational sector where I work, how the kind of coaching I'm talking about is different from other types of interventions. One of the beauties of coaching is that it creates the environment in which people can learn.

> But another way of defining it is as a one-to-one conversation, the purpose of which is to achieve a change or work towards a desirable outcome. So maybe coaching is a conversation that's designed to lead to change, or to help people to move towards more desired outcomes. I like the simplicity of it because in my view the role of the coach is to bring a simple structure and to focus on the condition, and the reason for that is because human beings bring the complexity, right? So each human is complex and what they're thinking about is usually complex, otherwise they wouldn't be bringing it to coaching. And, therefore, the coach doesn't want to add to that complexity by bringing in very complex theories or ideas.

> The process shouldn't be complex so that's why processes such as the grow model work very clearly: the coach's role is to just help follow that process through. I also think maybe the coach's role is to get rid of distractions and additional complications; bringing some clarity and the kind of attention into personal relationships that create a space for quiet reflection. That's slow thinking.

I asked Kim the same question.

> It's so many things now. I think this is probably how I tend to answer. Under
> the umbrella of coaching sit many, many different approaches, just like therapy.
> Coaching can be completely non-interventionist – it can be very akin to
> therapy – or it can be very provocative and playful, or it can be what's now
> being termed 'coach-sultancy', which is a blend of advising people what to do
> and using a questioning, coaching approach. For me (and I'm going to borrow
> a phrase from the Executive Coach Tom Preston), to begin with, coaching is the
> most fascinating conversation about you that you have ever had ('you' being
> the client rather than the coach). I think it's a confidential, focused, guided
> conversation where the client has complete freedom to go where they want to
> go, explore their thoughts, their feelings, their hopes, their dreams and their
> ambitions in the presence of an interested, thinking partner who will give them
> real attention and do their best to support and challenge and tease and provoke
> and stimulate their thinking so that they get the outcome they want. The coach
> will use any number of different interventions, tools or techniques.
>
> It's like working with a self-help book or personal development book which has
> come to life in the form of a person. One thing that I think is really important –
> and again I draw this from my therapeutic background – is that a coach isn't
> just a passive participant. How the coach is, the presence of the coach and
> the coach's own personal qualities play a large part in the development of
> the clients and their thinking. A lot of the time this means that the coach is
> emanating such acceptance of that other person that they are either explicitly
> or implicitly saying to the client that they are already OK as they are. And
> therefore here in this room, here in this space, you don't need to do what you
> might have done for the rest of your life, you don't need to be defensive, or to
> joke, or to get me to like you or believe you, or all those coping strategies that
> we all adopt because here, for the first time, I'm telling you that you're OK as
> you are just being you. That's probably for me the most important and subtle
> and underestimated part of the coaching process. A fixation on processes and
> goals can take coaches away from the core work. I certainly see some coaches
> having a bigger focus on the process and the outcome and the attainment of
> the goal and less on the psychodynamic aspects of the relationship and of the
> person.

Kim expanded her thinking to discuss coaching approaches:

> In a way, both approaches are OK. Transactional, short-term or behavioural
> attainment approaches – for example 'I want to do a marathon' or 'By this

time next year I'll become leader of my team' – are absolutely fine as 'Let's do this thing!' goals. They may have, sitting underneath them, behaviours or coping strategies which are going to derail that person throughout their life. In a year's time that person could be back to square one, presenting whatever they presented. So it's that whole dynamic between transformational change, which I think has to involve belief work, and transactional change, which doesn't. I don't think it's always appropriate to do both, but that should always be negotiated in the contract. Sometimes it would be very evident to me that belief work and transformational change at the deeper level probably need to happen, and in some cases I think coaches stay away from that and just deal with surface-level stuff. I've worked alongside a style consultant for many years and often seen her work wonders with women who lacked confidence. They have their hair done, make-up done, are dressed in wonderful new clothes and for a really short time they feel great. Then a few months later she would be back in her jeans with her top pulled over her bottom because she doesn't like herself, because actually it didn't address the core underlying issue of low self-esteem. At the same time I've worked with people sometimes who have low self-esteem and improved and addressed that by working at the deeper level, but looking at them you wouldn't know that because they haven't manifested that change in the outward behaviour. So, in a really long-winded way, what I'm saying is you usually need to do the transactional and the transformational, the behavioural and the emotional, together.

From Synaptic Potential's work with individuals and organizations we see this also to be the case. The 'all bases' approach we have pioneered is based on the blended approaches of neuroscientists and behaviour change experts. Just changing minds and then going back into the old environments can derail efforts. Vice versa can be equally unhelpful. While sometimes one approach or the other can suffice it is safer to address both wherever possible.

I asked Kim, as the head of a well-respected coach training organization, what she sees coaches struggle with most.

It's a really interesting question that nobody has even asked me. A few things come into my mind. One is accepting that they can do this work with the information and knowledge we give them. They don't actually need to go and do a seven-year PhD in some psychological discipline. I think they mystify therapy particularly, so one of the things I wrestle with is giving them the confidence to know that you don't need to be frightened in this work – it is a process. Coaches often say things to me like 'I'll be opening a can of worms or causing someone damage or starting something that I can't finish.' I'm perhaps a bit intolerant if they fear that, because I have an enduring belief in people's robustness.

This fearfulness is probably what leads some coaches to just stick with the transactional goal setting stuff, and I need to think about what more I can do as an educator to reassure them that they're not going to do damage. I like some of the provocative therapy models around that – the notion that the client has more of an ability to self-manage than perhaps the client knows. Because if the coach is thinking 'Oh gosh, this is dangerous, I think I'm going to damage them', that could lead to transmission in some way.

When coaching first started, it was very much about not looking at the past or people's personal history, but instead focusing on the goals. It was just about solution, focus, outcome. The world of therapy at that point was saying to coaches 'You can't do what we do because it's taken years of training to do this, and as long as we both keep separate it will be OK.' But as soon as people started showing up for coaching with all that they are, coaching realized that you have to delve into people's psychology sometimes and you can't always look forward. So coaches have moved closer towards the therapeutic model. Therapies have moved closer to a coaching model and distinctions between coaching and therapy are now harder to define. What provides the rigour for me in a therapeutic process is the work that the therapists do on themselves, and knowing how much they are an instrument of the change process. I'm not sure coaching has entirely caught up with that necessity.

I asked Kim, 'In an ideal world, what do you think might develop in the field of coaching?'

I think it will be an even greater requirement for supervision and professionalization and all the things that come with it. Any requirement for working in supervision so that coaches continue to work on themselves all the time, and perhaps a shift in emphasis in training away from tools and techniques and more towards the dynamics of the relationship. The aim is to put your ego aside and if you need to deliver some honest feedback to the client, just do it. If you feel like you are not having a good session yourself and you're not giving the client what they want to see, put everything out there for the client, and put your own needs aside. It's so much about the coach knowing where they stand, how comfortable they are with emotion, including possible negative emotion, how comfortable with conflicts, how much they are aware of their own prejudices and of conscious and unconscious biases.

There is a book by Irvin D Yalom called *Love's Executioner and Other Tales of Psychotherapy*. He is a psychotherapist who writes in a self-disclosing way about what's actually happening in his mind when he's working with someone, and I remember one story particularly about an obese woman. Yalom reflects in the book on his own feelings and biases about the woman's obesity and the challenges he faces when working with her. I think it's that kind of real, honest self-awareness that the coach needs to have.

In this section we are looking at classic areas of coaching, but it is useful to be aware that the field of coaching is continuing to develop. It is evolving as you coach.

Mindframes

'What are mindframes?'

What are mindframes? Well imagine for a moment that you come up with a great new idea at work. You stop to chat with your manager about it, and either intentionally, or perhaps without even really thinking about it, you will frame that idea in a positive way to try to persuade them of its merits. In doing so, you will sway their opinion in one way or another. Mindframing is based on the same principles but here you aren't framing an idea, you are framing the way you think about yourself or about the world around you. And the person you are trying to sway is yourself.

The term mindframes comes from the idea that there are many different frames that we can look through to filter information. The more common, similar, term is mindsets. However, we prefer mindframes because of the greater flexibility frames give you over something being set. There are some common mindsets that are useful for all coaches. Believing in the huge potential of each human being. Knowing people are capable of great change. Trusting individuals know what is best for them and have a wealth of answers within them. But, there are also different mindframes that are useful for different outcomes.

In this section we explore classic coaching concepts that are linked to particular mindframes:

- Self-control/willpower – this draws both on mindframes and skills. Our state, self-talk and beliefs all impact self-control.
- Optimism – this is more of a classic mindframe. There are times when optimism is the best mindframe and others when pessimism is useful.
- Mindfulness – viewing experiences through a mindfulness frame is very helpful to help yourself be present, experience it more deeply, form memories and allow reflections to bubble up.

- Expectations – these themselves are individual frames that information then goes through to be processed. Our expectations are hugely influential on our neurobiology and subsequent behaviours and feelings.

- Beliefs – beliefs are partly formed from the way that we interpret and explain things to ourselves.

Martin Seligman describes 3Ps that are also important here. Personalization (whether we interpret something as being from within us, or attribute it to outside factors). Permanence (how long we believe something will last – when is it useful to frame something as temporary and when more permanent?). Pervasiveness (how far you cascade a thought over into other areas of your life). Rather than thinking in absolutes, we encourage coaches and managers to embrace mental flexibility. Sometimes a person can set themselves up unhelpfully with framing things as permanent, solely on them and then look for all the ways that challenge shows up in their life. For example with failures… someone might see failures as permanent, as spanning more than just the event itself, and assign them blame solely to themselves – something that is likely to be more harmful to their wellbeing and performance. In another situation you might want to see a one-off success as a sign of permanence (that it shows what you can do and feel confident it will be one of many), that you're responsible for it (personalization) and pervasive (looking for other examples in life where you have achieved things you've set your mind to).

Self-control/ willpower

What is it?

Self-control or the utilization of willpower involves the ability to consciously decide what you will do and when, to resist temptation, and to put off a reward until later.

Why is it important for coaches/managers?

A person's ability to control themselves is key to their success. There are many mindframes that can support or detract from self-control and the construct of many goals involves the exercising of it. Sometimes just getting started on a goal requires willpower. For example, for someone who wants to run a marathon but is not that keen on running, they may need to exercise self-control to make themselves go for a run on their planned days rather than go out for a drink with colleagues. Once they have been on a few runs and experienced the flood of endorphins, and felt great as a result, they may find themselves keener to go out for a run than to do other things. Less self-control will be required. The inner dialogue remains important. Especially for logging the memories long term. Coaches and managers have an opportunity to helpfully draw attention to behaviours that serve people well. Good parents do it all the time. 'Well done for doing that homework even though you didn't feel like it,' or 'That was really kind to include Daisy in your game, even though I know she can be hard to play with sometimes.'

There are other goals that may require self-control throughout. For example, attempting to address a habit where someone regularly interrupts colleagues can be a constant challenge. It will take willpower and self-control to mentally and physically hit pause before the interruption has occurred and the damage to the relationship has been done. Understanding the latest findings on self-control can help you to set up your client with the best chance of succeeding.

A lack of self-control can predispose the client to create excuses for lack of achievement. If a person repeatedly does not achieve something, then rather than take responsibility for the reality of the situation it can be natural for some people to start making excuses. In the long term this can damage a person's chances of achieving future goals. Addressing any challenges around self-control in a safe and calm environment is a great investment for anyone you are working with.

What is the underpinning neuroscience?

Research has shown that exercising and depleting willpower causes a slow-down in the anterior cingulate cortex. You might remember from Chapter 6, where we looked at the anterior cingulate cortex, that this area is vital to self-control. It is commonly known for its conflict-monitoring activities. It is the area that picks up any mismatches between what you intend to do and what you actually do. (All those times you've put the toothpaste in the fridge while multitasking making your morning coffee... your ACC has picked up on it! It knows when you are making an error in the moment.)

The disappointing news is that a review of the scientific literature showed that there were strong behavioural indicators when willpower was depleted, but weak feeling indicators. This makes it slightly trickier for your client to know when they are about to snap at a colleague, for example, or lapse back into smoking.

The prefrontal cortex is also very important in controlling ourselves. The PFC has vital roles in planning, our working memory, organization, anticipation of consequences and controlling impulses. Often considered at the other end of the spectrum is the amygdala, which has a great deal of power over our decision making. Sometimes there is conflict between the rational cognitive parts of the brain and the emotive reactive parts.

What are any interesting studies in this area?

One of the pivotal studies in the field of delayed gratification is the famous Stanford Marshmallow Experiment (Mischel *et al*, 1970). This involved a series of children being sat in a room with a marshmallow on the desk in front of them. They were told that they could eat the marshmallow at any time or wait for the experimenter to return and then they would receive an additional marshmallow to enjoy. They found that most of the children couldn't delay their gratification long enough, although watching them try is both illuminating and comical.

Where the implications develop are when Mischel followed up many of the initial subjects many years later. His daughters, who were in the same school as lots of the children, would often tell him stories about how their classmates were doing. This prompted him to investigate further. He found that the children who were able to resist the marshmallow longest were achieving better grades, were more popular, eventually earned higher salaries and had a lower body mass index.

Brain differences have been found between those who can resist the marshmallow and those who give in to temptation. The key differences were found in the prefrontal cortex and ventral striatum. The study looked at 59 of the adults who had been in the original marshmallow experiment as children and represented people from both extremes of the ability to delay gratification (so the really good people and the really poor people). The nature of the test was changed to a more adult version, so participants were asked to signal when they saw a happy or frightened face. Dr Casey from the study explained: 'In this test, a happy face took the place of the marshmallow. The positive social cue interfered with the low delayer's ability to suppress his or her actions' (Casey *et al*, 2011). So when individuals with lower abilities to delay gratification saw the happy face they pressed the button when they shouldn't. The emotional connection of a happy face interfered with the rational instruction of when to push the button.

When the participants performed this test in the fMRI scanner the results showed that in the people who were able to delay gratification for longer the prefrontal cortex was more active. In those who struggled to delay gratification it was the ventral striatum that was more active.

Roy Baumeister has performed many interesting studies. One involved fasting college students. These hungry people were first left in a room with cookies, chocolates and radishes. Some were invited to eat the yummy things. Next they were given geometry puzzles to try to complete, although they were insolvable. (How long it takes them to give up is a standard technique to measure perseverance.)

The students who had been allowed to indulge in the cookies and chocolates kept working at the puzzles for around 20 minutes. This was the same as a control group (hungry but not tempted). The group only allowed to eat the radishes gave up after eight minutes. This tipped off the researchers to considering that willpower got depleted. This has big implications for how we optimally organize ourselves.

Another experiment required some subjects to suppress their emotions and others to amplify them while watching a sad film. The control group just watched it normally. Their stamina afterwards was tested (using the standard hand exerciser squeeze test) and unsurprisingly it showed that suppressing and amplifying emotions resulted in lower stamina.

In other studies Baumeister and colleagues (1998) suggest that blood sugar levels may drive the strength of willpower. This controversial idea states that the lower your blood glucose level the more likely your willpower is to fail you. Several of the studies in this series involved individuals doing lots of effortful cognitive tasks. Their willpower is then tested in some way, and they tend to fail. Those who are given a sugary lemon drink tend to have stronger willpower than those who get the sugar substitute in their lemon drink. This may link to the study above, where the students who ate cookies spent longer on the puzzles.

Carol Dweck *et al* (2013) have proposed that rather than the key being sugar, it is in fact our mindset that has a key impact on our willpower. They suggest that if we believe sugar will help then it probably will. Researchers at Stanford University explored people's beliefs about willpower and tested whether and when these individuals experienced a dip. They found that those who believed willpower was abundant didn't seem to need sugar in order to persevere through challenging tasks. Baumeister has also shown that people in good moods have stronger willpower.

We can often learn things about willpower from people who are unwell. It would be interesting to explore, for example, how people with anorexia nervosa appear to exhibit such strong willpower around resisting something as key to humans as food. Their sugar levels are low, and yet something enables them to retain a high level of self-control. By understanding this mechanism better it could provide additional help to people experiencing weak or exceptionally strong willpower when either of these is unhelpful to the individual.

These insights bring us back to a place of hope. We have hope that neuroscience will drill down into a more detailed understanding of how willpower works so that we have a more thorough understanding of how to improve it. Trial and error, as is often the case in such psychological studies as those we have covered here, can only take us so far. At times we are seeing seemingly conflicting information but the drive for a deeper understanding is spurring on every discipline involved in exploring this field.

How is it useful to managers who coach?

There are two critical things for managers to be aware of. Firstly, an exhausted or mentally depleted colleague will likely have their self-control compromised. Think about when people lose their temper. When people behave in meetings in less than helpful ways. The likely scenario is that the individual is tired, mentally overloaded or has been making too many decisions!

The second important concept is that the research tells us that self-control is easier when you are free to experience natural emotions. Part of your job is to create a safe environment where people can be themselves. If people are having to inhibit their emotional responses or pretend to be someone they are not, it will drain them and have a knock-on negative impact.

A final thought is around team meetings and snacks. Do you have team members who are trying to lose some Covid Kilos? Or just trying to be healthier? If there is an assortment of sugary snacks around then they may be investing precious energy inhibiting the desire to scoff them all.

What can I do with a client or colleague this afternoon?

Work with your client on ways to strengthen their mind's 'breaking' abilities and help them to come up with some creative ways to distract themselves. An example of this would be if someone was working on getting a promotion and had been told that they needed to speak out more during meetings. Perhaps the natural tendency of your client is to sit quietly in meetings. Internally they are having a great dialogue... but no one else gets to hear it. What might come into their mind just before they are about to speak is thoughts of what others may respond with, or questions that people may ask. This may result in an ongoing internal dialogue until their point no longer fits the external conversation that has continued. Maybe your client could experiment with having the numbers one to five written on a piece of paper in front of them during meetings. When a comment comes into their head they make a one-word note next to number one and then speak their comment out loud. This could act as a physical distraction and mind-breaking opportunity. Their aim during each meeting could be to get to five comments.

Suggest that your client commits to doing one thing at a time that requires willpower, ideally when they have not got lots of other cognitively draining things also going on. So if they were considering embarking on a new and challenging goal then check in on what else is going on in life at this time too. If you are a manager and someone wants to start a new project that will involve them working with a colleague where there has been conflict in the past, then careful self-control management would increase the chances of success. If this person is working on several challenging things at once then they may find it harder to exercise self-control.

Perhaps task your client with making a to-do list in order to get things out of their brain. This could reduce the cognitive load they are carrying around and might enable them to exercise more willpower when they need it.

Optimism 26

What is it?

Optimism is about having a positive attitude towards the future. There are two main types of optimism that are of interest to us. The first is blind optimism. This is the classic 'everything will be great' response, even when you have been gaining weight for the last two months and your clothes no longer fit – a scenario where this type of optimism hasn't been shown to be overly helpful. The second type is called realistic optimism. This is more the attitude of: 'If I make some changes then things will be great, making those changes may be difficult, but in the long run it will be worth it.' Ultimately, people who have a realistically optimistic approach to life tend to believe that they will be able to cope with the challenging or bad things that will happen.

Why is it important for coaches/managers?

- The field of personal development *can* be very pro rose-tinted glasses and so it is important to understand what this means.
- How we interpret life events can affect our life's trajectory.
- Equipping clients to be mentally robust and successful.
- Being subject to the optimism bias can have positive or negative effects – and being aware of these increases choice.

What is the underpinning neuroscience?

It is fairly widely accepted that generally we see different levels of activity in the left and right brain hemispheres (this is called cerebral asymmetry) when people are in positive versus negative states. This can be seen when we use electroencephalography (EEG) with people. It has also been shown that subliminal biases in attention and interpretation can be linked to an increased vulnerability to stress. We know that the left inferior frontal gyrus responds to good news and the right inferior frontal gyrus responds to bad news.

It has been shown through fMRI scans that when we think about upcoming positive future events this activates both our rostral anterior cingulate cortex and the right amygdala. It is possible that we are helped to imagine these future events by assessing emotions from past events. It is also thought that these two brain areas work together to downplay negative emotional responses. This could help us to be optimistic and driven to achieve our next set of goals.

What are any interesting studies in this area?

There are many fascinating studies on this topic. What follows are a couple of these studies along with an insight into the 'optimism bias'. Suzanne Segerstrom (2007) has done a particular experiment that looks at the behaviour of optimistic people and how it differs from pessimists. A group of people are divided into the optimists and the pessimists. They are then all given a variety of different anagrams, including an impossible one. The findings show that the optimists spend more time trying to solve it. It is hypothesized that this trait of being persistent stands the optimists in good stead for life, enabling them to achieve more – simply through being committed to doing more. Obviously persistence is a good trait only under certain circumstances. For example, for the entrepreneur who keeps knocking on doors, such as Colonel Sanders or J K Rowling, persistence is what enabled them to finally succeed. However, for firefighters who thought if they persisted for just a few more seconds, despite their instincts or their facts telling them otherwise, the outcome has proved to be very different.

Colin MacLeod *et al* (2002) from the University of Western Australia performed a very insightful study that could prove very useful to coaches who want to help people change subliminal habits. MacLeod's experiment tried to train students to either notice or ignore threatening words. They were shown both a nasty word and a neutral word at the same time on a screen. A split second later they were shown a target letter and they had to respond as quickly as possible by pressing a button. The experiment was set up so that for half of the students the letter always appeared in the space of the threatening word and for the other half in the place of the neutral word.

These students were then given anagrams to solve, some of which were impossible (a trusted way of stressing people!). Those who had been trained to notice threat reported feeling more stressed during this process. This is an indicator that focusing on negative rather than neutral things could affect our daily activities. If any of your clients need to reduce their stress levels then try looking at their anchors. You may have encountered anchors as a cognitive bias that predisposes us to rely heavily on existing information. Like the ship's device to keep us secure to one place, an anchor can tie us to thoughts, actions or even decisions too. Things may be masquerading as pessimism that is instead learnt responses.

A practical demonstration of a positive application of this work involved a group of students who were relocating. This could have extended applications to any organizations that routinely relocate employees. In this group of students, half of them received training to help them avoid threat and this half showed less stress and anxiety when they moved from Singapore to Australia. The classic sentiment of 'we get what we focus on' comes to mind here.

The optimism bias

Tali Sharot (2012) has researched extensively what is referred to as 'the optimism bias'. In essence, this is the tendency that most healthy individuals display to presume that bad things won't happen to them and that good things will. So people tend to underestimate the likelihood that they will get divorced, suffer from cancer or be involved in a car accident. It also affects our ability to predict things such as how long a project will take (ever find yourself taking longer than you thought over things?) and even how much we will enjoy a holiday.

People used to think that having low expectations was the key to happiness. This hasn't been shown to be the case. In fact a study by George Lowenstein (1987) showed that anticipation is powerful. He asked a group of students how much they would pay to receive a kiss from a celebrity over different time spans (so, for a kiss right now, in one day, three days, one year, etc). The results were surprising: the maximum amounts fetched were actually for in three days' time, not right now. So the pleasure of anticipation was valuable to people. Interestingly, people seem to prefer Fridays to Sundays (a day of work rather than a day of rest or play). Could this be linked to the absence of the good anticipation of tomorrow, the dreaded Monday?

Optimism has been shown to make you try harder. Interestingly, when people are given new information, such as the statistics of how many marriages end in divorce, not everyone updates their beliefs to reflect this data. Research has found that people adjust more in response to positive information, so if a person thought they had a 60 per cent chance of getting divorced, and then found out it was only a 40 per cent chance, they would make a bigger adjustment than if they believed they were only 10 per cent likely to divorce and found out the statistics pointed towards a 40 per cent chance. At these times, the frontal lobes are not effectively coding errors in predictions. Optimism appears to reduce our ability to update ourselves based on undesirable information regarding the future.

Is optimism good for us?

People who don't have positive expectations about the future are more likely to have symptoms of mild depression and anxiety. Subsequently, some people say that optimism is vital to our mental health. Some studies also point to the benefits to our physical health, which include living longer and more healthily. Even in those people who do get sick, such as with cancer or AIDS, the survival time of these people who are also optimists is greater than the pessimists.

Optimists enjoy reduced stress levels (which causes an overactivation of the autonomic nervous system and hypothalamic-pituitary-adrenocortical axis). They catch fewer infections and have stronger immune systems. It has also been suggested that this positive outlook promotes healthier behaviour, so exercise and healthier food choices. Even in the professional world optimists come out on top with higher pay than others.

Downsides

Optimism can pose risks, including people behaving in ways that are not as precautionary as would be good for them. For example, having unsafe sex, not attending medical screenings, not having vaccinations or being adequately insured. Extreme optimists are more likely than mild optimists to smoke and fail to put aside savings.

How is it useful to managers who coach?

Consider the culture you are investing in. Broadly speaking, an optimistic vibe will serve your people well. They will try hard and persevere in this kind of environment. But managers need to be switched onto when this may be unhelpful. Can you guide people in putting on their 'black hat' (from Edward De Bono's famous 6 thinking hats book) to be critical and explore things from a pessimistic perspective at times?

In your position of seeing your team interact regularly, hearing their ideas, seeing how they respond to things you can help individuals to raise their self-awareness. Some people are less aware of how they are filtering information. The opportunity for group coaching and group reflection is huge.

What can I do with a client or colleague this afternoon?

- Explain what optimism is and what the benefits and risks with it are.
- Support them to brainstorm anywhere that they are unhelpfully exhibiting the optimism bias.
- Help your client to connect to the pleasures of anticipation. Suggest they make a list of everything that is happening this month that they are looking forward to. Perhaps each night they can choose one of these things to do a couple of minutes of mental simulation around, imagining what it will be like.

If your clients identify themselves as pessimists and express a wish to change, then things that may help are:

- Mindfulness-based therapy – these courses are meditation-based and normally run for around eight weeks. Several centres are now offering them in the United Kingdom and the United States and in other parts of the world they are even more prominent.
- Cognitive-bias modification – there are many ways to do this now available in the marketplace.
- The practice of 'an attitude of gratitude' where you suggest your client does a variation of finishing the day with three things they are grateful for – an exercise that has roots in this way of thinking.

Mindfulness 27

What is it?

Mindfulness meditation can be thought of as the practice of attending to present moment experiences and allowing any emotions and thoughts to pass without judgement. It can involve activities such as focusing on specific physical sensations – breathing, for example, which one observes as thoughts, emotions and bodily sensations rise up and then pass away. A newer, two-component model of mindfulness has also been put forward. This involves both the regulation of attention on immediate experience and also approaching experiences (regardless of what they are) with a mindset of curiosity, openness and acceptance.

Mindfulness has some historical links to spiritual practices. Buddhists practise mindfulness and it has been linked to achieving enduring happiness. There are many good books available specifically on mindfulness so here we will focus on the neuroscience component.

Why is it important for coaches/managers?

Being able to be present and 'in the moment' is seen as a desirable trait. In the past we might have thought that some people were just naturally more able to be that way, luckily having a more 'composed' or 'calm' personality. Now we know that it is possible to change our brain connectivity and rewire ourselves to support present-moment awareness. Mindfulness has been shown to elevate positive emotions, to have stress-reducing effects and pain-management effects.

Mindfulness meditation works in several ways:

- attention regulation;
- body awareness;
- emotion regulation;
- change in perspective on oneself.

Coaches may find that they are working with clients on goals or outcomes that a mindfulness meditation habit would support the client in achieving. Overall it appears that anyone wanting to use their brain for cognitively challenging tasks (so anyone working in a knowledge-based role) could benefit from mindfulness.

Mindfulness is well established as a psychotherapeutic intervention for anxiety, depression, substance abuse, eating disorders and chronic pain. It has also been shown to help improve immune function, reduce blood pressure and cortisol levels and increase telomerase activity.

What is the underpinning neuroscience?

In general, evidence shows that changes occur in the anterior cingulate cortex, insula, temporo-parietal junction, fronto-limbic network and default mode network structures. Specifically, it appears that mindfulness-based meditation leads to less involvement of the dorsomedial prefrontal cortex (DMPFC) during interoceptive attention (IA). Drawing on some of what we learnt in Chapter 4 about the insula we can now add that mindfulness also leads to altered functional connectivity between the DMPFC and the posterior insula. It leads to increased IA-type activity in anterior dysgranular insula regions, which is consistent with integration of interoceptive sensations with external contexts – or, in other words, appropriate feelings. Overall, we're seeing that this type of interoceptive training modulates task-specific cortical recruitment.

Table 27.1 lays things out clearly. Each of the mechanisms we understand to be key in mindfulness have been seen to correspond with specific areas of the brain.

Table 27.1 Key mechanisms in mindfulness as they correspond with specific brain areas

Mechanism	Brain area
Attention regulation	Anterior cingulate cortex
Body awareness	Insula, temporo-parietal junction
Emotional regulation – reappraisal	Dorsal prefrontal cortex
Emotional regulation – exposure, extinction and reconsolidation	Ventro-medial prefrontal cortex, hippocampus, amygdala
Change in perspective on oneself	Medial prefrontal cortex, posterior cingulate cortex, insula, temporo-parietal junction

Long-term meditation has been shown to increase cortical thickness in anterior regions of the brain (including the medial PFC and superior frontal cortex). Conversely, reduced cortical thickness has been found in posterior regions of the brain (including inferior parietal cortex and posterior cingulate cortex). In short, people who have meditated for a long time have structurally different brains to those who haven't.

Being able to focus attention is a very important skill for anyone working in a role where they have to use their brain to think a lot. We know that the anterior cingulate cortex enables this executive attention to take place by acting as a conflict detector. It detects the conflicts between incoming information streams that are being processed and that are incompatible with the task goal. It alerts other parts of the brain to regulate these streams.

Additionally, recent studies have indicated that different types of meditation involve different brain networks or neuroplasticity. For example, the Theravada traditional meditation that focuses on loving kindness has correlated with increased grey matter in the parahippocampal gyri (for five-year-plus meditators). This area of the brain is involved in cognitive empathy, anxiety and mood.

What are any interesting studies in this area?

Many studies have been done over recent years and this has become a very hot topic (see, for example, Lazar *et al*, 2005; Lutz *et al*, 2008). One study looked at whether mindfulness practice had any impact on functional connectivity between default mode network (DMN) regions (the specific set of regions that are engaged when people are left to think to themselves undisturbed, eg remembering the past or envisioning the future). The participants of the study went into an fMRI scanner during a resting state and the data collected showed that experienced meditators, when compared with beginner meditators, had weaker functional connectivity between DMN regions involved in self-referential processing and emotional appraisal. They had increased connectivity between some other regions, notably the dorsomedial prefrontal cortex and right inferior parietal lobule. The conclusions that can be drawn from this is that meditation appears to change functional connectivity in core DMN regions and that this could reflect strengthened present-moment awareness.

A study led by Dickenson (2013) looked at the neural mechanisms of focused attention (in this case the practice of focusing on one's breath).

The participants had to complete a task while utilizing mindful attention (focused breathing) and control circumstances (unfocused attention) while undergoing an fMRI scan. It was observed that during focused breathing a neural attention network was recruited that included parietal and prefrontal structures. This leads us to believe that attention processes benefit from trait mindfulness. Tang and Posner (2009) showed that five days of integrative body–mind training may lead to greater activation of the rostral ACC during the resting state. Tang (2010) also showed that after 11 hours of this training there was an increase in white matter integrity in the ACC. Other studies have shown that this initial enhanced activation of ACC may later decrease as expertise grows and the ability to focus attention is steady. Electroencephalogram (EEG) data showed increased frontal midline theta rhythms during meditation. This form of brainwave is associated with attention-demanding tasks.

Several studies have documented the stress-reduction effects of mindfulness:

- Brown and Ryan (2003) showed a reduction in mood disturbance and stress following mindfulness interventions.

- Jain *et al* (2007) highlighted the specific effectiveness of mindfulness versus other relaxation techniques. They suggested that it had a 'unique mechanism by which mindfulness meditation reduces distress'.

- Jha *et al* (2010) found that meditation training practice could offer protective benefits against high-stress environments.

A specific mindfulness-based stress-reduction programme was created by Jon Kabat-Zinn and the use of this is increasing worldwide.

Studies have also shown that:

- Mindfulness-based interventions (MBI) are effective for reducing depressive symptoms.

- Mindfulness-based stress reduction can reduce emotional reactivity.

- MBI may help people to stop smoking.

- MBI can help executive control, eg emotional acceptance and performance monitoring.

Dispositional mindfulness is an innate characteristic that reflects individuals' naturally occurring ability to inhabit this intentional stance of awareness. Studies have found that people with higher dispositional mindfulness (measured by the Mindful Attention Awareness Scale) predicted increased activation in the ventromedial PFC, medial PFC and ventrolateral PFC. It also predicted reduced amygdala activity and a stronger inhibitory association between the amygdala and PFC.

How is it useful to managers who coach?

One of the opportunities of being a manager is that you can introduce people to new practices in a non-threatening way. It is encouraging when people share that they have invested further in a practice that they first learnt about at work. It might be something they start using with their kids, or their partners or just by themselves. Mindfulness lends itself wonderfully to this. Imagine what positive impact you could have on your people's wellbeing if you started each meeting with a 60 second meditation. Nothing fancy (although it may develop into a light guided meditation). Many organizations do this, and their people love it. Even those who were sceptical at first, accept it and agree it makes them feel calmer at least.

What can I do with a client or colleague this afternoon?

Introducing your clients to mindfulness could benefit them in many ways:

- more success at work – increased ability to focus;
- better health – less stress;
- improved relationships – better emotional control.

We know that people who are mindful enjoy these benefits and we also know that mindfulness appears to be able to be learnt. Referring your clients to a course or book on mindfulness could be something they thank you for. Then for clients who choose to pursue an increase in their meditative capacity, offer support through your coaching programme with them.

Mindfulness may even be something that if you are not yet practising yourself, you choose to look into.

The five facets of mindfulness are:

1 Observing (noticing internal and external stimuli such as sights, sounds, smells, emotions, sensations and thoughts).

2 Describing (giving these stimuli a name – such as 'fear' rather than 'Bob').

3 Acting with awareness (being cognisant of what you are doing rather than allowing your mind to be elsewhere).

4 Non-judgement (not giving a positive or negative value to anything that is happening).

5 Non-reactivity (allowing everything to pass by rather than attention getting stuck in individual thoughts or feelings).

Body scan meditation

This is a simple practice that you could try yourself:

1 Lie down comfortably on your back. Ensure you won't be disturbed and are warm enough. You can either close your eyes or keep them open, or open them during the meditation at any time (especially if you feel yourself falling asleep!).

2 Bring your awareness to your body and the physical sensations you can perceive. Focus on the points where your body is connected with whatever you are lying on. Each time you breath out, feel yourself sinking deeper into that surface.

3 Consider this as an opportunity to 'fall awake', becoming acutely aware of what you are actually experiencing, rather than whatever you think should be. During this time you are not *trying* to feel or be anything, your only task is to increase your awareness of what is. Equally, if you are not feeling anything at any point, physically or otherwise, then that too is absolutely fine and only to be noticed.

4 Take your awareness first to your tummy, noticing the changes as you breath in and out, and any areas that feel different to others.

5 Next gather your attention into a spotlight with two beams and travel it down to your legs, into your feet and out into each toe. Notice what you feel – contact, tingling, warmth, numbness, nothing? Whatever you do or do not feel is whatever it is. There is no need for any judgement to be placed on anything you feel. You can allow any of the sensations to just be whatever they are.

6 As you next breathe in, imagine that inhalation going through your lungs all the way down your body and out into your toes. When breathing out, imagine that breath flowing out of your toes, your feet, your legs, tummy and finally out of your nose. This is a different experience for many people so may take a little playful practice.

7   When you are ready to move your attention (on an out breath) from your toes, bring it to the sensations in the bottom of your feet. Notice each of the different parts of your feet; explore the sensations present there.

8   Allow your awareness to fill the rest of your feet and then, on an out breath, let your feet go and move the attention to your lower legs.

9   As the name implies, you can continue to scan each part of your body one section at a time. Perhaps you might spend 20–30 seconds on each area, maybe a little more or a little less. The method is just to keep your attention there for as long as it feels right and then move on, rather than counting breaths or using a stopwatch!

10   If you find any areas of intense sensation, for example an area of tension or pain, explore breathing into this area. Breathing in, bringing your attention into the sensation and breathing out, noticing if anything changes.

11   Finally, experience the body as a whole, breathing in and out, holding as many of the sensations in your awareness.

Mindfulness exercise

Another straightforward exercise to try is to practise doing any normal activity mindfully. The activity can be anything – cleaning your teeth, doing the washing up, driving a short journey, getting dressed, or anything you do regularly. The aim is not to enjoy the activity more or to do it in a different way. The only aim is to be as *awake* or *alive* to it as possible. Focus your attention on each of the sensations you experience as you do it. Notice any thoughts or feelings that you have as you do it.

More on mindfulness from Susan

For her, Susan said, coaching is fundamentally always about enabling people to be able to make decisions for themselves and to feel more resourceful when they come up against potential challenges or decisions. People go to her for coaching when they're at some sort of crossroads and they've got decisions to make. She sees her role as helping people to see either which path forward is the right one for them or to see the path they haven't seen yet.

> I'm thinking of a couple of clients I'm working with at the moment. Something they both said right at the start of the relationship was how overwhelmed they felt. I think that's really common. People feel overwhelmed with information, with choices, with the pace of things. Where I think mindfulness comes in is if you're at that crossroads and you're thinking, 'Where do I go from here?' and you're feeling overwhelmed, that creates a bit of paralysis. So often people will do anything or perhaps they become a bit impulsive. Mindfulness helps people to manage that overwhelm, if that's the right way to put it, to recognize that that's what's happening, that they're feeling overwhelmed. Elements of neuroscience underpin that. I don't always explicitly explain the neuroscience, but that's what is in my mind when I'm helping people understand why they are finding it difficult to make decisions and the impact the overwhelm is having on their ability to move forward.

Susan has dug deeper into mindfulness than most people. She completed a Master's in mindfulness from Aberdeen University, so she has both thought about it and practised it herself and with others. Through mindfulness Susan helps people experience more. Initially she may help them increase their awareness of what overwhelm, for example, actually feels like. Where do you feel it in your body, and what thoughts do you notice? She explains that mindfulness is a particular type of awareness about your experience. Susan often helps people understand their experience more. She uses a coaching version of Daniel Siegel's hand/brain model that we teach on our programmes:

> People just get that. They get it so quickly. I use that model often for people in the situation when they feel like they don't know which way to go. They've got all these options and they can't sleep at night because of all these thoughts going through their head. I help them understand what is actually going on. I can talk about the neural networks and the executive functions of the brain and the impact that the primitive parts of the brain can have over that executive functioning.

> We talk about how the brain processes information but also how you can bring that mindful awareness and your ability to focus on your brain together. It all helps to reduce and manage overwhelm. I don't see neuroscience and mindfulness as distinct elements of what I do: they seem to merge together. Maybe it is indicative of the way that I've processed what I've learned from

neuroscience and what I've experienced with mindfulness. For me it all weaves in together, and that is really helpful in serving my clients.

I asked Susan what she had noticed broadly across her work with mindfulness: 'You've mentioned to me previously that most of us are in our heads a lot of the time and detached almost from the body – what kind of impact or problems are you seeing this create?' She replied:

> The biggest problem with this, which is a macro-level global problem, is that people are experiencing stress, mental health issues and other types of health issues. They are not noticing the signs their body is giving them. They push too hard or they're not dealing with particular emotions. They are basically not admitting to themselves that there's potentially something they need to focus on because we're so fast-paced: 'I've got a bit of a niggle, but you know what, I'll just push through it.' And that can be a niggle in terms of physical discomfort, it could be a mental thing, it could be a thought that keeps recurring but we just try to push through. I have situations with my clients where they've ended up burnt out or developing some significant mental health issues. I notice through the mindfulness course I run that although there are people who have been diagnosed with anxiety, other people in the course had never been diagnosed, had never been to the doctor, but by doing things like a body scan, being aware of their breath and noticing that actually they are anxious quite a lot of the time, they realized they had recurring headaches. It's really important to notice those things because our bodies are really quite wise and generally speaking, if there's something that keeps coming up and we push it away, then it's likely to cause more problems in the future. Do we want to go through our lives feeling below par? I meet a lot of people who don't feel quite as good as they could do. If you connect with what your body is telling you, then you've got a chance to do something to adjust the balance. Coaching gives people a space to recognize balance. Mindfulness gives people a way to deal with that balance. Neuroscience then comes in because you can talk about how the different parts of the brain work with each other and how one bit can impact the other area of the brain, so again for me it all kind of weaves together.

Susan and I are like-minded in that we believe that you can best serve your clients when you have awareness from multiple disciplines. Essentially you underpin your approach with in-depth understanding from multiple camps. Neuroscience is always a great lens to view things through and a great knowledge to have because almost every other discipline also uses the brain.

We also both love a good evidence base. In fact Susan described what happened when she first came across mindfulness:

It was the evidence base that initially hooked me. I'm not naturally somebody who would buy into something that seemed so simple and almost too good to be true. I questioned how it could be possible. Surely people are being duped by this? But the evidence base is massive and it's growing daily.

Some people teach mindfulness as a quick fix to everything. Susan and I both see this and it frustrates us because it doesn't end up serving people. She says:

> Some people talk about mindfulness almost implying that if you just sit for 10 minutes and focus on your breath, everything else will be great. Actually there's a lot more to it than that and you do need to practise it. It's like any skill or any approach: you have to know what to do and you need to practise. Some organizations 'tried mindfulness' and made a ruling. I think credible mindfulness practitioners help people see that there could be a benefit to investing time. This is where I think the evidence base really comes in. So what does sitting focusing on your breath actually do? Well, it soothes you, yes, but actually what we know is it helps you train that muscle of attention in your brain so you can then focus on other activities for longer. It helps to calm that fight-or-flight response that goes off in your brain. So I think neuroscience is an evidence base for mindfulness and helps people to see that investing time is worthwhile.

In many ways coaching and mindfulness seem to go hand in hand – but I was curious about Susan's thoughts and experiences. When I asked her: 'Do you think the process of coaching can be done unmindfully as well as mindfully?' she replied: 'My instinct is to say yes, it probably can. I think if you are experiencing really good quality coaching I think it's probably evidently mindful. But I think coaching can sometimes be unmindful.'

I continued: 'Linking to your earlier point about just being in your head and rational, then certainly it's possible to coach in that style, isn't it?'

> Yes, one of the main contradictions between coaching and mindfulness is coaching, generally speaking, is very goal-oriented and very future-oriented, whereas mindfulness is very present-focused and not goal-oriented. I run a mindfulness and coaching programme, and that's something that we talk about. There is a potential contradiction between being mindful and coaching but I think there's definitely a way that you can do both because mindful doesn't mean you never have goals. It's about how you approach those goals. If you become very fixated and rigid on that goal then there is a danger you become a little blinkered to what else is going on. So I think you can mindfully seek to achieve a goal, which means 'Well, I know where I am trying to get to.' I

favour the word 'intention', so I often talk about setting an intention rather than goal. Goal also has that feeling of achievement or non-achievement, and there's a bit of striving that comes in. And again that's not mindful. Mindfulness is not striving to achieve something. So if I don't achieve, then what? Whereas mindfulness is about: If I set an intention, for myself, for my business, or for my organization then I can still move towards that in a goal-oriented way, but I'm maintaining a wider awareness, so for me that would be a mindful approach to achieving goals. I think my approach to coaching has shifted with my experience with mindfulness, so when we talk about goals it's much less 'Let's have a smart goal and we must look to achieve that.' It's a bit freer and a bit less constraining.

You will notice parallels with this shifted approach to both Christian and Kim and I'm sure many others. I think it is a beautiful example of how coaching has evolved. I asked Susan to contrast a coaching session that started with 'Right, let's start setting SMART goals' with how she approaches things.

Typically the very first thing I do is a five-minute... I don't call it a meditation, but I just say 'OK, let's just sit for a few minutes.' Then I talk them through feeling their body sitting in a seat and focusing on their breath, because I like to start my sessions with people feeling grounded, a bit more relaxed and less in their head. I always say to people, 'If you're quite happy let's just sit for a few minutes and leave all the stuff outside the room so we can totally focus.' I haven't had anyone yet who says 'No, I don't want to do that.' I haven't yet had anyone who's looked at me funny. People tend to say: 'Do you know, I actually feel more relaxed.' Then I say to people, 'OK, so what is your intention? What are you wanting to achieve from the coaching?' I also use more open questions than some coaches. I prepare to go wider than just 'What's your goal, your timelines?' So the first thing is getting people in that grounded space. The second is to be wider in terms of what your intention is. And the other thing I do quite frequently occurs when we're looking at the actions – the 'what are you going to do?' part of coaching. Here I take people back to 'Let's just sit for a minute. Close your eyes. So here's what I've heard you saying', and I'll play back to what they've said their goals and their actions are. When I say to them, 'Just tell me what you experience, what are the feelings you get in your body when you hear this played back to you, when you hear these actions.' That can be really quite fascinating. For example, last week I was working with a client via Skype and she said the process was really interesting because she got a sudden really obvious knot in her stomach. It followed this flow: 'So you said this was the goal that we had talked about,' and she said, 'Yes, that's it, I am really clear that's what I want to do,' but yet when she came out of her head back into her body, her body was saying something about it to her through this node in her stomach. We explored that a little bit, and it came down to: she didn't actually believe that she would be able to do it, so part of her – her brain – was saying, 'Yes, that makes sense. You should do that and that will all work out.' But something at subconscious level was saying to her, 'You don't really believe this is possible, do you?' You can get there with coaching without doing any sort of meditation or anything like that. But I try to get the people to connect with their beliefs and the subconscious elements of what makes their decision making more quickly. That's why I do another very practical thing. I mentioned compassion earlier. I think when we get into a very goal-focused, striving way of thinking, we can actually sometimes be very unkind to ourselves. If we're not achieving what we've set out to achieve, this internal critic who is very loud tells us we're not good and everyone else is doing better than us. We can be quite unkind to ourselves and not very compassionate. So another thing I bring in to coaching is helping people to recognize that sometimes external factors will get in the way. It doesn't mean that you failed. You can still be focused on your intention, but give yourself a bit of a break and just be a little kinder to yourself.

I asked Susan what the future holds for mindfulness.

More and more organizations are becoming interested in it and I'm having more and more conversations with organizations who are saying, 'I heard about this mindfulness thing. What is it? What can it do for us, and how can we introduce it?' I think if it's positioned in the right way we could easily see mindfulness being a key part in a lot of organizations. Some have mindfulness rooms or meditation rooms. Some are using things like one of the many apps out there. Some colleagues get together as a group and spend time listening to the meditations together. I think how people are mindful at work will evolve. I would love to see mindfulness become more integrated into the culture of an organization. Coaching is definitely getting there. We don't just coach individual managers; we look at the culture that they're working in and how that needs to change too. An organization can run mindfulness courses for people and you can learn and experience and practise and enjoy the benefits. The concern is that sometimes those people go back to an organization where the culture is almost the anti-physics of mindfulness. Things can be so fast-paced that people don't have the opportunity to pause and reflect. Another important facet of mindfulness is compassion: considering how we communicate with each other and how we support each other. I think a lot of organizations, particularly when there are tough times economically, miss out on compassion. It can be quite difficult to continue to be mindful when you're in an environment that's not really supporting that, so I think that's an opportunity as well as a challenge for organizations.

I did a small piece of research earlier this year, and even within four weeks the managers I was working with said that mindfulness was a really important part of how they dealt with the challenges they faced at work. They said that the external stuff which caused them stress and overwhelm was still there, but now they were practising mindfulness they could approach those things and deal with them differently and more effectively. And that's just four weeks, so I think that's fantastic. But again from my research what became evident was the long-term impact of that maybe impeded by the fact that the culture of the organization was changing, so I just wonder, for those managers back in the organization, how much influence they can have in changing the culture? Possibly they can, but they have to be strong and resilient to stay mindful in a very unmindful culture. I think that's a big challenge.

Expectations 28

What are they?

An expectation is what we think or hope is most likely to happen. We have expectations all the time; it helps us manage our world. When an expectation is unrealistic, though, there is a high risk that it will not be met.

Why are they important for coaches/ managers?

Everyone has expectations that they may or may not be consciously aware of. When expectations are not met then it can have an adverse effect on a person. When they are met or exceeded it can have a very positive effect. Understanding these and how to use them to your advantage can be useful. In the workplace, often the expectations of others also need to be managed for a smooth relationship.

What is the underpinning neuroscience?

Expecting something positive to happen can have the effect of reducing anxiety. This can be useful if, for example, a person has a meeting coming up and they want to feel calm about it. They may want to imagine it going well and them conveying their points effectively during this meeting. It can also activate the brain circuits that are involved in reward mechanisms. These reward circuits link together both cognitive and emotional responses, and even motor responses. This may mean that someone enjoys the benefits of a reward circuit activation by just thinking about a positive expectation they have.

On the contrary, having a negative expectation can increase anxiety and activate the areas of the brain involved with negative experiences. This obviously affects our state and subsequently has the potential to affect the end result.

So, when we expect a primary reward, dopamine cells seated in the nucleus accumbens fire off in anticipation. When we are not expecting a reward but receive one (such as a surprise compliment, a hug, finding a £5 note on the floor, etc) we get an even bigger dopamine hit. When our expectations are unmet – like when we don't meet a deadline, gain weight we'd expected to lose, don't get that promotion we'd been promised – the fall in dopamine levels is painful. These and a huge number of other unmet expectations can also cause a mild threat response.

As you'll remember, dopamine is really important for us in order to be open, curious and interested. How fast we can think is increased with more dopamine – too much, though, and our minds become chaotic. Being able to hold an idea and focus in on something is dependent on the right levels of dopamine.

What are any interesting studies in this area?

Scott *et al* (2007) did an imaging study that looked at the role of the nucleus accumbens and expectations. They took subjects and used functional molecular imaging to observe the dopamine release when people were being given a placebo. They found that the activation was related to the anticipated effects of the placebo! They then did some functional MRI studies, which looked at the expectation of monetary gain. Here they found that this expectation increased the nucleus accumbens synaptic activity proportional to the dopamine release when people were given placebos.

Another study, this time by Volkow *et al* (2004), looked at the effect of expectations and the reinforcing effects they could have. In this study, cocaine users were split into four groups:

1 People expecting a placebo and given a placebo.
2 People expecting a placebo and receiving the drug methylphenidate (some similarities to cocaine).
3 People expecting the drug and receiving the drug.
4 People expecting the drug and receiving the placebo.

The results were significant. When people expected the drug their brain glucose metabolism was 50 per cent higher than when they didn't expect the drug. People also self-reported feeling high by 50 per cent more.

When people are given two identical bottles of wine, with wildly different price tags, they tend to say that the more expensive one tastes better! (This gives us good reason to leave the price tag on a good bottle!)

Expectations have also been shown to have a powerful relationship with pain and emotion. We have seen expectancies affect the processing of pain in the insula, cingulate cortex and thalamus. In placebo studies, if we expect something that we are swallowing to reduce our pain, it often does. In another study (for which the details can be confusing, so here we will focus on the results) researchers discovered that our powerful cognitive bias reflects a general property of sensory information processing. In the experiment they test whether the contents of visual awareness could be altered by manipulating expectations. The effects were dramatic and showed that expectations have a strong influence on our experience of sensory input.

How is it useful to managers who coach?

A critical area that impacts performance is 'performance management'. Performance reviews, KPIs, goal setting. This whole package can help or hinder people's ability to do their best. At Synaptic Potential we've consulted to several companies on this challenging topic and managers have a pivotal role to play. People need to own their 'targets'. They need to believe they are achievable.

Another area of work that is linked to this concept is remuneration/rewards. This is a huge topic, beyond the scope of this chapter. But we need more champions in organizations who understand how brains work so can promote wise policies. We want people to do their best. We want them to achieve what they intend to. Routinely, this should be the norm. People should be rewarded for doing what they set out to do. It is unwise to encourage people to repeatedly try to do more than they have logically identified is sensible. Highly rewarding people for overachieving won't get the best result in the majority of people over time.

What can I do with a client or colleague this afternoon?

- Work with your clients to evaluate their expectations and whether they are serving them or hindering them. Mindfulness practices may help some clients to become more aware of their expectations.

- If you coach leaders then help them to examine what expectations their teams have and what effect that may be having on them.

- A classic situation in many business environments is having unrealistic targets that switch people off before they even begin – help clients to address this if it is an issue for them.

- Ensure that clients have planned things to look forward to.

Beliefs 29

What are they?

Beliefs are tricky from a neuroscience perspective. When we look at commonly accepted definitions we see things like: 'A belief is an acceptance that something exists or is true, especially one without proof.' Or 'trust, faith or confidence in something'. We have a gut feel for what our beliefs are and what we would say are facts that we hold to be true. We can recognize that different people hold different beliefs. In coaching we often say that beliefs can be powerful and have significant effects on people.

Angel and Seitz in 2020 recognized that the topic of beliefs has been neglected in the natural sciences. They suggest though that they are the neuropsychic product of fundamental brain processes. These attribute affective meaning to things and events which helps with goal setting, decision making and navigating through life. The fact that beliefs are, essentially, created has great importance to coaches and managers. Eliminative materialism, supported by neurophilosophers Paul and Patricia Churchland, suggests that our common-sense understanding of things, such as beliefs, will one day be proved false. In simple terms, they do not believe that beliefs exist! Instead they propose that neuroscience will uncover a deeper level of understanding that renders the concept of beliefs no longer useful. B F Skinner, a radical behaviourist, was also a supporter of this way of thinking and likened pseudoscientific theories (eg the four humours and phlogiston theory of combustion) to mental states like beliefs.

The Churchlands argue that folk psychology should be evaluated based on its ability to predict and explain things in researching the mind and brain. Neuroscience is able to do this and can explain many cognitive processes that folk psychology is still unable to. Daniel Dennett and Lynne Rudder Baker, who suggest that while beliefs are not a scientifically valid concept they can be useful as a predictive device, propose a more middle ground. If we play chess with a computer, most of us don't think that the computer *believes* that by taking our queen it will have a good advantage. However, we would be sensible to act as if that were the case because it will likely inform the moves that the computer makes. Dennett suggests that beliefs occupy a different level of explanation to those of fundamental neuroscience. Perhaps this is a little like trying to explain a number using colour.

Why are they important for coaches/ managers?

Many coaches and managers (sometimes without realizing) work with their client's beliefs – and so understanding what they really are, how they are formed, how new ones can be formed and being open to the future offerings of other constructs may be useful to you.

Sachs and Hirsch suggested in 2008 that:

1 humans tend to believe their perceptions to be true;

2 humans develop a positive stance towards their beliefs;

3 beliefs can be updated by confirming or disproving new evidence;

4 the processes underlying believing result from brain function.

These are important implications for us to be aware of.

What is the underpinning neuroscience?

Beliefs are relatively poorly defined from a scientific perspective. This causes some problems when trying to identify the underpinning neuroscience. What follows is a range of different approaches that have been explored. Damasio gives us some insight into the complexity of the task when he states: 'it is likely that the mechanisms which allow us to develop the basis of beliefs, as well as the mechanisms by which we retrieve and express them, are operated in a largely covert manner' (Damasio, 2000).

Often in science we learn more about what is normal by examining the abnormal. In this case many people have explored delusions, an impaired perception or belief, in order to better understand what constitutes a normal one and how it may form. However, delusions themselves are also poorly defined. The American Psychiatric Association's handbook (1994) states that a delusion is:

> A false belief based on incorrect inference about external reality that is firmly sustained despite what almost everybody else believes and despite what constitutes incontrovertible and obvious proof or evidence to the contrary. The belief is not one ordinarily accepted by other members of the person's culture or subculture (eg, it is not an article of religious faith).

There have been some correlations between specific delusions and anatomical findings, for example in sufferers of Capgras delusion in which people believe that someone close to them has been replaced by an imposter. A significant proportion of Capgras delusion sufferers have been found to have lesions in the right hemispheres of their brains. Unfortunately, research in this area is not yet producing findings clear enough to be of real use to coaches.

Kathleen Taylor from Oxford University suggests that beliefs are like memories from a neural standpoint. She has written a book that looks at the extreme end of beliefs: brainwashing (Taylor, 2006). Taylor looks first at the basics of neurological pathways, including the communication of neurons and how pathways get grooved in, realizing that the more something is repeated the stronger the neural pathway becomes, which we know is summarized in Hebb's law. When people have strong beliefs it is likely that the neural pathways are strong, perhaps from repetition or intensity. From looking at the techniques employed by cults Taylor shares insights into the importance of repetition and the emotional pull of ideas and new beliefs.

The concept of beliefs is alien to children until they are around four or five years old. If you tell a three-year-old child a story about Jack and his magic bean that he put on the kitchen table, but his mother then moved to the porch, and then ask the child where Jack *believes* the bean to be... most will reply on the porch.

A study that looked at the brains of committed Christians and non-believers gives us some insight into beliefs. The individuals were positioned in functional magnetic resonance imaging (fMRI) scanners and then asked to evaluate the truth and falsity of religious and non-religious propositions. What they found was that when people believed something there was activation in the ventromedial prefrontal cortex. We know this area is linked to self-representation, emotion, reward and goal-driven behaviour. This activation was not dependent on whether the thing that was being believed was a 'Christian' fact (such as the Virgin Birth, for example) or an 'ordinary' fact. When there was a comparison made, they observed that the 'ordinary' facts were more reliant upon memory retrieval networks than the religious beliefs.

What are any interesting studies in this area?

One interesting study that has been linked to beliefs was carried out by Matthew Lieberman *et al* (2005). The study involved both African-American and Caucasian-American individuals who were asked to lie inside fMRI scanners; they were then shown three images of African-American and Caucasian-American people on a screen (one at the top and two below). They had to match the race of the person at the top to one of the two at the bottom. They also had to complete an experiment where they were given just one face and two written words, one of which described the race of the person in the picture (either Caucasian or African-American).

The study was designed to look at any differences between amygdala responses in African-American and Caucasian-American individuals to African-American and Caucasian-American faces. The amygdala is known for its activation to threatening, novel or highly arousing stimulus. It has been implicated in race-related processes too. The results of the experiment showed that both African-Americans and Caucasian-Americans produced greater right amygdala activation to the African-American target than to the Caucasian-American. Previous studies only tested Caucasian-Americans, which meant that the reason for any amygdala activation could not be attributed to a potential threat or the novelty of the African-American face. By this study including African-Americans as test subjects their brain responses can be compared. The authors of the study note that this is not conclusive, although suggest that the observed amygdala activity may be a 'reflection of culturally learned negative associations regarding African-American individuals'.

The other interesting finding supported other studies in the difference between perceptual encoding and verbal labelling effects on the brain. Here, when people saw the words rather than the pictures their right ventro lateral prefrontal cortex was activated. The theory is that this inhibits the amygdala. Some beliefs will light up certain areas of the brain. Many of our beliefs (if we use this term) are picked up unconsciously and we may not even be aware of some that are having an effect on us.

Joel Winston *et al* (2002) carried out another study that asked people to rate the trustworthiness of each of the faces they were presented with. We can think of this as: 'How trustworthy do you believe this person is?' Winston found that the amygdalas in the subjects responded most to the pictures of the faces that had been chosen to represent untrustworthiness. People's ratings didn't always correspond with their brain response. Does this mean that we should trust their brain or their verbal account? What implications does this have for how we work with people?

How is it useful to managers who coach?

We are often encouraged to accept people's beliefs. In many contexts this is vital and right. However, when we are in the role of being a thinking partner and trying to help a person to unlock their potential and become the best version of themselves possible it becomes important to be able to challenge beliefs in a supportive way and in a safe environment.

To be clear, this isn't about imposing one person's beliefs on another. This is about exploring together what the most useful beliefs may be for each individual. For example, at times it may be helpful for one person to believe something negative. It could motivate them to change a situation. However, for another person, that same belief could be demotivating and negatively impact their mental wellbeing.

What can I do with a client or colleague this afternoon?

The suggestions here are mostly familiar to coaches already. Unfortunately at this point we are waiting for science to move on a little more in order to see whether beliefs as a construct will survive another 10 years or whether our understanding will develop. It might be that as a coach you choose to look at and work with the neuroscience in the meantime. Here are some considerations:

- Acting as if beliefs are real:
  - Explain at least the neural mechanism of how thoughts, ideas and possibly beliefs get strengthened (through repetition and emotional intensity and the relevance of Hebb's law).

  – False beliefs are possible in individuals. Therefore, a worthwhile exercise could be to challenge and explore any potential false beliefs.

- The alternative, bold opportunity is to explore how things would be if beliefs did not exist:

  – Explore with your client what they know to be true and how they know it to be true.

  – Explore what your client may now be free to do or be.

SKILLS

To be an amazing coach there are many skills that can help; however, there are only a few that are vital (when combined with a belief that individuals have a huge amount of potential that they can access). The training team for both Neuroscience for Coaches and Coach Savvy Manager have seen the same skills be used in very different ways. To do transactional coaching, which focuses on a specific problem or goal. It stays shallow, doesn't dig into a person's identity. Or to coach the person and get into deeper self-awareness and patterns of behaviour, thought and what makes a person them.

Managers often feel limited to stay in problem-based coaching. Many even use the term as a smoke-screen. Believing they are coaching when they are asking a few questions but blatantly just telling people what to do. They are not showing up as thinking partners. It is possible to coach more deeply as a manager. With solid frames and tight self-control for the manager it can be hugely rewarding for individuals. Managers see their team members frequently in a work context, so if they can have these colleagues' best interests at heart, create a safe space and strong relationship then they can have valuable observations and reflection to share.

The foundational four: safe space, listening, questioning, reflecting

This set of skills are included only for a sense of completeness. They are truly foundational and are covered in most training programmes. I won't shoehorn some neuroscience behind them as the mechanisms are multiple and complex and some (like creating a safe space) are covered elsewhere.

What is it?

There are at least four foundational skills that every coach and manager benefit from investing in. The first is being good at creating a safe space for a person to open up and think and share. The second is to listen well. Third is the skill of questioning, which is more than just asking a range of different types of questions: it is important to know which questions to ask when. The fourth is the skill of reflection, which is to offer a mirror back to people to help them see their mindframes, behaviours and skills more clearly.

Why is it important for coaches/managers?

Coaches are typically taught that they need to build rapport with their clients. That they need to establish confidentiality boundaries. Sometimes that is as far as it goes though. Understanding just how vital a safe space is for higher level processing to take place helps both coaches and managers search for more ways to continuously establish safety. It is important to remember that, while you continuously need to build it, it can be damaged quickly and easily.

Listening is important for multiple reasons. The effect of someone feeling heard is huge. When experienced regularly, it can reprogramme someone's social stress responses. Another clear benefit of properly listening is that you can be a proper thinking partner. It links with the next skill of asking good questions. A question can, on paper look brilliant, but if it is poorly timed or misses something of the context then it can be unhelpful. So proper listening also facilitates questions that lead to insight or progress.

Finally, the skill of reflection is important to help really coach the person not just a goal or problem. In this privileged role, when you have a safe environment, you can offer observations for the person to consider. It may be they disagree which is fine, but they may find value in an additional perspective. Ideally, they will become more self-aware and seek out other routes to increasing their field of view.

Flow

What is it?

Being in flow can be described as a specific optimal state. It is a psychological construct and is often called being 'in the zone' or 'on fire'. It can be characterized as engaging in challenges that stretch you, but that you are capable of. Tackling a series of goals, continuously processing feedback about progress and adjusting action based on this feedback are key ways that the state of flow is achieved. Psychologist Mihaly Csikszentmihalyi proposed this state and suggests it is possible to experience it in any area of life: work, relationships, hobbies, sport etc.

Why is it important for coaches/managers?

The state of flow is a highly regarded construct. It is covered here because even if further research uncovers multiple constructs in the future (as opposed to just the state of flow), the 'knowledge to action' gap means that the concept is likely to be around for some years yet. Flow is considered productive, fulfilling, satisfying and positively feeds back into you.

The state of flow has the following characteristics:

- intense, focused concentration on the present task;
- action and awareness are merged;
- low self-consciousness;
- sense of autonomy;
- feeling as if time has passed quickly;
- activity is intrinsically rewarding.

When people are either overstimulated or understimulated their levels of performance are negatively affected. To perform well requires the 'Goldilocks' level of arousal: enough to be energized and focused but not too much so that anxiety and panic freeze your cognitive abilities and ability to act. Today, more so than ever, organizations and cultures have been lean-ing towards rewarding and validating people who work like crazy. The irony is that it *is* crazy! Pushing and rushing is not conducive to the way the brain works optimally. Too much arousal can lead to high levels of cortisol – and over time this can damage our brains.

What is the underpinning neuroscience?

Currently the neural correlates of flow are not conclusively researched (I'm hopeful that this will change in the near future... certainly researchers realize that there is a gap here and more studies are being carried out even at going to print). It appears that theta activities in the frontal lobes are higher during flow when compared with being overloaded or bored. This may be linked to a high level of cognitive control being required and immersion in the task. Some research has found that the middle temporal gyrus (BA39) is the delta wave generator and the medial frontal gyrus (BA10) is the alpha and beta generator. These findings support the idea that flow requires a transient hypo-frontality, theoretically to momentarily suppress the analytical meta-conscious abilities of the explicit system, in favour of the implicit system mainly localized in the basal ganglia circuit. It has also been suggested that rapid attentional shifting between salient cognitive precepts correlates to feelings of elation and satisfaction.

The neurochemical process that enables this shifting increases cognitive efficiency and creativity. Cognitive shifting has been associated with dopa-mine release, and subsequently with reinforcement. As you will remember, dopamine enables us to give our attention to something (it stabilizes active neural representations in the prefrontal cortex); dopamine can give us ad-vanced rewards before a behaviour has even occurred. The way that people describe the state of flow could be linked to the cognitive effects of dopa-mine. It is conceivable that mesolimbic dopamine facilitates switching be-tween alternative cognitive perspectives and so enhances decision making and creative thinking. Neuroimaging studies have shown an increase in do-pamine release during activities that require sustained shifting of a cognitive set, such as playing video games. It has also been suggested the process of flow involves the chemicals norepinephrine, anandamide, serotonin and endorphins.

There appear to be strong links between the practice of mindfulness and the state of flow; more research is needed, however, to flesh out the details. We do know that neuroplasticity enables us to rewire our brains. It is plausible, although untested, that some brains are currently better wired than others to facilitate the state of flow. For example, some people's brains may be better wired for focusing. When we are focusing well we tend to enjoy what we are doing more and don't experience undesirable ruminating and inefficient distractions. We end up performing better – quicker and with fewer mistakes – when we are focused.

What are any interesting studies in this area?

There is an interesting study by John Pates that looked at the effect of hypnosis on flow states and the performance of basketball players (Pates, Cummings and Maynard, 2002). It was suggested that people enter an altered state of awareness during hypnosis and move from a cognitive and analytical processing state to a more holistic and imagined one. There is some evidence that this is the case and that EEG activity shifts from the left to right hemisphere during hypnosis. Plausible next-stage suggestions are that hypnosis assists in purposeful hemispheric shifts that are conducive to performance, eg control over flow. Pates did a couple of very interesting things that are useful for helping people to access the state of flow. The first is to use trigger-control techniques: anything that can act as a trigger (such as words, sounds, images, components of tasks). The second thing was to encourage, in this case basketball players, to recall and submerge themselves in the memory of a time where they performed optimally. Associating this state with a trigger enabled them to access this optimal performance during a future event. In this study, the players enhanced their three-point shooting performance. It showed that flow can be accessed using hypnotic regression and trigger-control techniques.

Some studies have shown that work environments should have open space, windows and enable people to work without being interrupted. There should be a culture of achievable challenge and a higher purpose in what you are doing. These things increase the likelihood of people getting into the state of flow. For the state of flow to occur, it is important that your skill level matches the challenge. You need to be capable of doing the task.

An experiment from a very different angle involves transcranial direct current stimulation (tDCS). In essence, this is a small electrical current that is directed through specific parts of the brain. Some studies have shown this to produce in individuals a state similar to flow. This electrical current is thought to depolarize the neuronal membranes in the area of the brain that it is being directed to. This makes the cells more excitable and responsive to inputs. It is possible that this process accelerates the formation of new neural pathways when an individual practises a skill. In one experiment using tDCS the subjects were tasked to play a computer game in which they were shooting at targets: the results were that the response speed increased by 2.3 times what it was without the tDCS. How tDCS works is not completely clear yet, but we know some interesting things. For example, tDCS gives people an advantage as soon as the current starts to flow; this leads us to believe that the mechanism is not solely about writing new memories, because the process wouldn't be that quick. A possible explanation links to the theory that activity in the prefrontal cortex is reduced. We know that this area is active during critical thinking, and Csikszentmihalyi has shared that this is reduced when people are in flow. Neuroscientist Roy Hamilton believes that tDCS may have much broader effects than we currently realize. While some neurons are muting signals of other brain cells in their network, it is conceivable that by stimulating one area of the brain, other areas may reduce their activity.

There are several studies that look at flow at work. One study looked at 258 secondary-school teachers. The researchers were exploring the relationship between personal and organizational resources and work-related flow. The hypothesis was that personal resources (for example, self-efficacy beliefs) and organizational resources (for example, social support and understanding clear goals) facilitate work-related flow and also the converse, that work-related flow has a positive influence on personal and organizational resources. Flow at work can be thought of as a short-term peak characterized by absorption, enjoyment and intrinsic motivation. An absorbed employee is in a state of total concentration and experiences time flying while they are immersed in work.

The researchers focused specifically on self-efficacy (our belief in our own ability to achieve something). There is a lot of research into self-efficacy already and it has been shown to be linked to resilience, an increased sense of competence to control and impact one's environment, and work engagement; it is seen to be a buffer to stress and is linked to better health, self-development and social integration. The researchers found that personal resources and organizational resources did indeed facilitate work-related flow.

The experiment made use of the self-assessment WOLF (work-related flow) scale. They also found a causal relationship between work-related flow and a positive effect on personal and organizational resources. Apparently, feeling competent in the present was a good predictor of being in flow in the future.

Another very different study looked not at the individuals and the state of flow but at flow experiences at the group level. It has been previously noted that *collective flow* is more likely to occur when a group is performing at the peak of its abilities. Even more interestingly, a multi-sample study showed that social flow is more enjoyable than the individual flow experience. This fits with other research that we are aware of around the importance of our social nature. According to social cognitive theory (SCT), when people collectively share the belief in their ability to achieve something this may impact how the group perceives challenges and may contribute to the experience of collective flow.

How is it useful to managers who coach?

Flow is becoming more of a mainstream skill for individuals and teams to tap into. It is proposed as a skill as it can take practice to both create the right environment to get into the state of flow, and to allow yourself to enter it. Managers can help or hinder this process.

- Ensure people have enough to work on that challenges them but also engages them and is within their capabilities.
- Help describe tasks in ways that draw out the opportunities that can be intrinsically rewarding.
- Take a big step back (big brother looking over a shoulder is an almost certain flow killer).
- Remove or reduce as many distractions as possible.
- Provide clear and regular feedback.
- Equip people to own the task or project.

What can I do with a client or colleague this afternoon?

- If your clients are not yet familiar with flow you could encourage them to read up on it.

- Ask them to keep a note during a normal day of how many times they think they were in the state of flow, and what specifically they were doing at that time.

- For clients who are not sure when they experience flow, coach them through discovering when they might be in that state (for example during sport or music) or strategize what they could try in order to experience it.

- Consider embarking on a programme addressing any self-efficacy beliefs and helping your client to build and strengthen these.

- Look for ways they can increase opportunities to get into the state of flow every week.

For clients who are struggling to find this state of flow, a more comprehensive approach may be beneficial. Perhaps start by sharing with them the analogy of thoughts being like trains. The trains may keep coming into the station, but you have the choice of whether or not to get on. Your thoughts may keep coming into your head, but you can decide whether or not to go with the thought or just let it pass out of the station again. Our thoughts can be incredibly distracting – and for the state of flow, like with mindfulness, it is important to be present. Strengthening our ability to let our thoughts pass and instead to stay in the moment could make it easier to get into and stay in the state of flow.

Motivation 32

What is it?

Motivation is what drives and sustains a person to act towards a goal. It has movement and direction. This goal may be a conscious and clearly laid out one or it may be more subtle and unconscious. Often the situation is such that motivation is only noticed when there are challenges with getting things done, and it is then that people want to address it. Motivation is a psychological construct, but from a neuroscience perspective it involves the dopamine reward circuitry in the brain. We are driven to undertake activities that are rewarding to us and avoid punishment to us. We form predictions about what will reward us and we use these to form the basis of decisions that guide our behaviour.

Why is it important for coaches/managers?

The ability of a client to achieve a goal is heavily dependent on how they manage their motivation to keep taking the steps required to move towards that goal. If a person is not motivated then they are unlikely to take enough action to make something into a habit or to get the result they want. Motivation is a key concept that coaches and managers need to be aware of. This is true for both people working with individuals on personal aspects of their life and for people working on work-related things.

Imagine a manager who, acting daily in the capacity of a coach, understands the science behind how people are motivated. The power that individual would have to help blend together the elements that would best serve their team would be immense. The knock-on benefits that the team would experience would be great. They are likely to enjoy increased productivity, improved job satisfaction, increased loyalty and a sense of belonging, among others.

Some people ask whether it is possible to motivate another person – and there are challenges inherent within that question. If we think of motivation purely in neurological terms then motivation exists within the individual. However, we are not just our brains – our environment and behaviours can and are shaped by others.

What is the underpinning neuroscience?

There is a lot to motivation, and there is not scope to cover everything here but we go into it in depth in both the Neurosciecne for Coaches programme and the Coach Savvy Manager programme. However, recent reviews of the scientific literature draw distinctions between two types of motivation. One focused on motivation and goal-directed behaviour, the other focused on habitual or stimulus-response learning. Let's start with the habitual or stimulus-response learning. At a simple level, motivation involves the reward circuits in the brain. Dopamine neurons in the mid-brain signal a reward prediction error, which lets us predict – and take actions to increase – the probability of receiving a (neurological) reward in the future. This is the underpinning to lots of habitual learning. For example, if your client does something at work and then their boss says 'well done, that piece of work really helped us secure important new business', then it is likely that your client will be more motivated to repeat the behaviours that led to the praise – and that gave them the dopamine hit.

The reward system in the brain has important links to the area of motivation. Rewards can reinforce behaviour. If we perform a behaviour that is then positively reinforced, the motivation to do it again increases. A reward is considered a reinforcer if it increases the probability of the recurrence of a behaviour. The reward circuits include the dopamine-containing neurons of the ventral tegmental area, the nucleus accumbens and part of the pre-frontal cortex.

The reward system uses a major neurochemical pathway, the mesolimbic pathway. This goes from the ventral tegmental area (VTA) via the medial forebrain bundle to the nucleus accumbens (NA) (within the ventral striatum). The star neurotransmitter of the reward system is dopamine and this is released primarily by the VTA. The way dopamine works in this reward system is not yet clear-cut. There are three solid hypotheses:

1 Hedonia – here dopamine is positioned as a 'pleasure neurotransmitter'. When people take certain addictive drugs and experience a 'high', studies have shown dopamine to be involved. However, it has also been found that not all rewards or pleasurable experiences involve dopamine.

2 Learning – in this context learning involves predicting future rewards and forming associations. Studies have shown that rats whose VTA and NA have been destroyed are still able to learn, but are not motivated to work for a reward.

3 Incentive salience – the feeling of wanting something may be a front-runner in the hypotheses. When we want something, dopamine is released in order to help us to work hard to get the thing.

Goal-directed behaviour is different, however, and the underpinnings here are a little less well researched. We do know that the prefrontal cortex is important in goal representation, decision and planning. A lot of the research has been done in rats and there is good evidence that reward learning, grooving in of stimulus-response habits, is under dopaminergic control. The processes going on here, though, seem to be neurally, behaviourally and computationally different from goal-directed actions. It is suggested that there are parallels between goal-directed behaviour and nondeclarative memory (unconscious memories, skills). It is not surprising that decisions that people make involve memories that they have, nor that these cognitive capacities are shared.

While the neuroscience is a little thin on the goal-directed behaviour front, we have a good body of knowledge of habitual actions and the motivation that is involved there. From a coaching perspective we do know that it is important to help clients to develop habits that support them, and memories that help to guide and drive them.

Making decisions

When we are considering our options we need to weigh up which will give us the biggest reward. For example, perhaps for the small-business owner it may be the choice between relaxing on the sofa or working on their new website; for the working parent perhaps it is getting an extra 30 minutes of sleep or getting up and going for a jog. Our brain is responsible for predicting where the highest reward will come from. This process is a complex one and involves balancing exploitation and exploration – exploiting what is already known and coded for in terms of rewards from actions, and exploring new actions to see what the reward is. As individuals our responsibility is to make it easy for our brain to wire in rewarding behaviour repeatedly, so that it makes it easier for us to choose it again in the future. As coaches and managers our responsibility is to support and guide people in doing this.

What are any interesting studies in this area?

We'll start with a psychology study led by Sam Glucksberg in 1962. Since the 1940s an experiment called the 'candle problem' had been used, which involves asking a person to fix a lit candle to a wall in a way that the candle wax doesn't drip on to the table. They can use anything in front of them, which includes a book of matches and a box of thumbtacks. To do the task the person has to overcome prior programming telling them that the box is for the thumbtacks. The answer to the problem is to tack the box to the wall and put the candle in the box. What Sam did that was new was to offer participants a financial incentive. One group were told that if they were quick then they would be given some money (enough money to be attractive at that time). What happened to the group who were offered this incentive, though, compared to the group who were not, was that they were *slower*. That's right... the financial incentive made people slower (Glucksberg, 1962). In fact, in such circumstances as this, the more money people are offered, the slower they become!

Another interesting study was done at MIT where they took groups of students and gave them challenges to carry out, including things like solving word puzzles, number puzzles, even physical things such as throwing balls (Pink, 2011). They incentivized the students using a three-tier system. The results were that with mechanical skills the bonuses did increase productivity: the students performed better. However, as soon as even rudimentary cognitive skills were required for tasks, the rewards scuppered people's ability to do them.

Reinforcement learning

One experiment involved a monkey that was irregularly given drops of juice that were signalled by prior visual cues. The researchers found that when the reward (the juice) is unexpected (in this instance if the visual cue is not reinforced or there isn't a cue) then dopamine neurons are excited. On the other hand, when the reward is expected because there has been a reliable cue, then the dopamine neurons do not respond. On a sliding scale, the more expected the reward, the less the neurons respond. When a reward is expected but it fails to materialize then the neurons are briefly inhibited below their baseline-firing rate. So this 'reward prediction error' shows that dopamine neurons fire relative to the baseline to notify the difference between expected and actual reward.

Error signals are useful to us to update our predictions about the future. Evolutionarily it has been advantageous to be efficient. If we can improve our ability to gauge the likelihood of a reward, then we can evaluate how much energy we want to expend on something. If something is not rewarding to us then we are not motivated to do it.

Reward

A study looked at the responses to reward and punishment in the basal ganglia. Participants in this study were put in fMRI scanners and then had to guess whether the value of a card they were about to be shown was higher or lower than five. They were given a financial reward if they guessed correctly. Different trials gave them reward (+$1), punishment (−50 cents) or neutral feedback. Parts of the dorsal and ventral striatum responded differently to this feedback. After a reward feedback there was sustained activation, whereas it decreased below the baseline after punishment feedback.

A review of findings from neuroimaging studies shows the orbitofrontal cortex to be involved in coding stimulus reward value and working with the amygdala and ventral striatum to represent predicted future reward. In one study, participants were scanned while being presented with predictive cues. These cues were one of two odours. One of the odours was then devalued in the brain of the participant by using selective satiation. This enabled the imagers to see that the brain regions that responded to predictive cues, tracking changes in reward value to the odours, were the orbitofrontal cortex, amygdala and ventral striatum.

Salience – how noticeable or important something is to us

The striatum is implicated in coding for how salient something is, similarly to dopamine. The striatum has been activated in studies where 'non-rewarding' salient events occur. When the involvement of dopamine neurons in rewarding versus salient events is debated then it has been useful to see whether these neurons also respond during salient but punishing events. A similar approach can be taken when looking at the involvement of the striatum and salience versus reward. Interestingly there are studies that show that the striatum *does* respond when an aversive stimuli is present, for example during pain or the anticipation of pain. When someone is expecting a reward, but doesn't get it, parts of the ventral striatum are deactivated. It is important to note that not getting an expected reward is equally if not more salient than getting an unexpected reward.

How is it useful to managers who coach?

Managers, it is important to pay attention to the effect that expectation has on the brain. We saw from studies that expecting a reward that then does not materialize has an effect. We can liken this to a demotivating effect. On the other hand, an unexpected reward can be likened to a motivating effect (it isn't quite as simple a translation but there is potential for the effect to be as such). Therefore, precautions need to be taken to ensure that people are best set up to take actions that trigger the reward system. This may be as simple as reviewing targets to be achievable or to have positive social interactions within teams. Alternatively, it may be more along the lines of you giving your team unexpected rewards – such as a surprise box of doughnuts, letting them go home early, or words of praise in front of other teams.

When we work with organizations on the challenges around motivation we almost always refocus onto engagement. When engagement is strong then problems of motivation melt away.

What can I do with a client or colleague this afternoon?

Help your client to identify what already motivates them. This is likely to be causing a release of dopamine. See if you can unpack any obvious links between what is driving them and what is important to them.

People generally respond well to:

- autonomy;
- purpose;
- mastery.

So in your coaching consider focusing in on each of these areas and exploring how your client is getting and working towards each of these things.

As an experiment, listening to your playlist on random or to your favourite radio station may give hits of dopamine as the songs play and surprise you.

Decision making 33

What is it?

Decision making is the process of selecting one choice from multiple options. You can decide to do something, think something or feel something. We make countless decisions every single day and it is only relatively recently that we have been able to look at how this process occurs in the brain.

Why is it important for coaches/managers?

People make decisions all the time, mostly unconsciously. Many of their results are the product of the decisions they make. Achieving goals hinges on people being able to make decisions that will lead to the accomplishment of those goals.

Lots of people still believe that decision making is a purely rational process. This gives coaches an opportunity to expand that understanding to take into account the most recent findings in this field.

What is the underpinning neuroscience?

Decision making is a complex cognitive task, which includes many variables. Making a decision can involve accessing memories, a value system, cognitive biases and an emotional response. Often we need to accumulate evidence for and against the different choices available to us, evaluate the possible outcomes and risks, and suppress learnt responses – and all these processes can be happening unconsciously.

Several parts of the brain have been identified as important to this complex process, including the orbitofrontal cortex (within the prefrontal cortex) and anterior cingulate. The orbitofrontal cortex is responsible for processing, evaluating and filtering information both of a social and emotional nature. The anterior cingulate is responsible for controlling and selecting appropriate behaviour and monitoring errors and incorrect responses. The dorsolateral prefrontal cortex (dlPFC) is involved indirectly in decision making. It is involved in working memory, planning and flexibility.

There are some other interesting things around decision making from a neuroscience perspective. Sometimes we selectively search for evidence of something and disregard evidence for something else (termed confirmation bias in psychology): for example, listing several people who are old and who have always smoked or eaten an unhealthy diet, while ignoring all those who die from diseases related to these activities. People who are very defensive in this kind of decision making have been shown to have increased activity in their left prefrontal cortex.

We can be primed to pay more attention to recent information when compared to things we haven't come across in a while. For example, if the last article you read on exercise endorsed the benefits of interval training then you may do that at the gym tonight rather than your usual distance training.

Only recently has it become possible to do single-neuron recordings that extend beyond sensory and motor coding, beyond what we see or do. It is now possible to study the processes involved in how we value and make choices. It is useful to remember that in the area of decision making our behaviour is very adaptive. Our environment is very important and this is reflected in studies that record the activity of single neurons.

What are any interesting studies in this area?

There are researchers seeking to understand the physical substrates of mental phenomena. Scientists are out there trying to work out how our thoughts are generated by brain processes. When dealing with something as potentially subjective as thoughts, a quantitative approach is valuable. This is something that economic models bring to the field of decision making.

There are a range of types of studies that look into decision making. There are multiple single-neuron recordings that identify which neurons are active during a decision-making task; there are PET scanning studies and functional MRI studies that show increased activity in bigger areas of the brain. The non-invasive scanning experiments are where we focus.

Antonio Damasio has been a pioneer in shaping our understanding of the role of emotions in decision making (Damasio, 1996). Damasio's somatic marker hypothesis suggests that the decision-making process is one that actually depends on emotion. Studies have shown that when the ventromedial prefrontal cortex (vmPFC) is damaged, and the normal emotional signals can't get through, advantageous decision making is disrupted. Other studies have shown that the amygdala also affects decision making but in a different way to the vmPFC.

Alan Sanfey has enlightened us greatly into social decision making by drawing on experiments from game theory (Sanfey, 2007), which is a collection of rigorous models that bring to light complex decision-making interactions. He explains that the field of neuroeconomics is integrating cognitive and neural constraints of decision making, bringing together psychology and neuroscience.

Studies using the trust game, prisoner's dilemma and the ultimatum game have all uncovered interesting insights into how our brains work when making decisions. Some of the themes that have come out of this research include social reward, competition, cooperation, coordination and strategic reasoning. It is believed that the mesolimbic dopamine system is involved. Changes in the striatum have been shown to correlate with increasing magnitude of monetary rewards or punishments. It has been suggested that the striatum may register social prediction errors to help guide decisions about reciprocity. One study showed that beliefs about the person they were about to play a game with reduced the amount of 'trial-by-trial' learning required, which showed that the neural basis of social cooperation goes both top-down and bottom-up.

An fMRI study looked at unfair behaviour in the ultimatum game and found that the anterior insula showed greater activation as the unfairness increased. This activation predicted whether or not the player would accept the other player's offer – what decision they would make.

Several studies have highlighted the difference between cultures where individuals tend to be more or less independent. This can have a big impact on decision making. For example, in North America people are more likely to decide to buy sun-tan lotion after watching a promotion-based advert, while on the other hand, East Asians will be more likely to buy sun-tan lotion after watching a prevention-based advert. This only holds true when the people are time-pressured, though: this means that if you are working with someone who has to make quick decisions then looking at what their biases are, and how they are predisposed to make decisions, could be useful to them.

Another interesting component to decision making is the phenomenon of *Schadenfreude*. This is where people 'delight in others' misfortune'. From a social comparison theory perspective, when people around us are in pain in some way (socially or otherwise) we look better to ourselves. In one MRI experiment that was designed to measure empathy they found that men (but not women) enjoyed seeing bad people suffer. The experiment involved participants watching 'bad' or 'good' people being given electric shocks. The researchers expected to see the empathy centre of the brain light up, which it did, but in men so did the pleasure centres when they thought that the subject was deserving of punishment.

There are links between *Schadenfreude* and envy: one study could predict the brain's *Schadenfreude* response based on the previous envy response; a person's ventral striatum (reward centre) was activated when they were told that people they envied had suffered misfortune.

How is it useful to managers who coach?

At Synaptic Potential we are frequently asked to train teams in high-level decision-making skills. Often with the added challenge of being time pressured. It is a critical skill for organizations. As a manager consider the following:

- How you can protect against biases – could you ask people who are different to you for their ideas? Could you take time to reflect on the decision? Sometimes could checklists help?

- How you are creating psychological safety – people can think better, and therefore come up with a better quality decision when they are mentally safe.

- How you frame people and the decision – remember the power of framing? Even the words you use to describe a required decision are influential on people's thinking.

What can I do with a client or colleague this afternoon?

Task clients with reflecting on some of their best decisions over the last year. Which ones do they feel really good about? Reverse engineer how they believe they made these decisions and look for any patterns. Are they aware of any intuitive decision-making component? Do they tend to need the logical arguments to stack up before they can commit to something? Have they ever had the experience of making a decision and then feeling like it was a good one once it was made?

Explore with a client the value of emotions in the decision-making process. Can they see where they may have made different decisions based on their emotional state in the past? Where has that helped or hindered?

TOOLS

As with skills… you could continue accruing tools for decades. However, there are some core ones that no coach should be without. A word of caution here: while tools can be used to really enhance coaching interactions, if they are used without the Foundational Four Skills then they can do more harm than good. In a coaching situation, even to work shallowly with a person, you are asking them to be vulnerable. To talk about a goal can be a revealing thing. To share habits that are hindering oneself can be tinged with shame or other challenging emotions.

We've frequently seen managers with only a little coach training be tempted to overly rely on tools. It can seem safer than having that open, honest, raw conversation. It can be tough to go with the flow more, but can also facilitate great thinking that leads to shifts and behaviour change.

Goals 34

What are they?

A goal is a desired result. Goals can be traditionally tangible or intangible. There are ways to measure both and they can both be brain friendly. Some researchers propose that the attainment of goals relies on three core processes: goal maintenance, performance monitoring and response inhibition. Goal maintenance is one's ability to keep a cognitive representation of that goal in mind. Performance monitoring is the ability to keep tabs on the end state and current state, and identify any lapses in progress. Response inhibition is about being able to stop those behaviours that aren't in alignment with moving towards the goal.

Why are they important for coaches/ managers?

- Goals are at the heart of coaching.
- Not achieving a goal can affect a client's mindset and even identity negatively.
- Repeatedly achieving goals can make it easier to continue doing so.

What is the underpinning neuroscience?

The neuroscience of goals is a huge field and we cover it in depth in our Neuroscience for Coaches programme. Our brains are clearly important in the setting, pursuing and achieving of goals. The scope of what is involved in achieving a goal is very wide. Achieving a goal through taking appropriate action has components associated with it:

- deciding on the goal;
- thinking creatively;

- planning how you will achieve it;
- visualizing the goal, and the steps to reach it;
- paying attention to/being affected by relevant cues – conscious and unconscious programming (anchoring, nudging, priming);
- exercising willpower or self-control;
- helpful habits;
- monitoring progress and adapting;
- inhibiting internal distractions (emotions);
- inhibiting external distractions;
- taking action.

Having the backing of neural networks set up to best support your clients and colleagues in achieving their goals is really important. These can be set up and strengthened in a variety of ways. Not paying attention to this core level of programming is setting people up to have more challenges than they need to.

The existing research points us in the direction of certain brain areas as being involved with goal-related processes in general. These are the dorsolateral prefrontal cortex (dlPFC), ventrolateral prefrontal cortex (vlPFC), dorsal anterior cingulate cortex (dACC) and parts of the striatum. When we get to the point of actually taking action it is the regions responsible for motor control that get involved. Namely these include the primary motor cortex, supplementary motor area, premotor cortex, cerebellum and basal ganglia. The basal ganglia seems to act as a motor control hub in integrating information from inputs and then outputting in response.

What are any interesting studies in this area?

There aren't many neurocognitive studies that address goal pursuit in its entirety, but several studies look at different components of it. It is quite difficult from a study design perspective to take into account the complexity of goal pursuit and to study this with tools like functional magnetic resonance imaging (fMRI). Here we consider several studies that look at individual components of working towards goals and then one that looks at the coming together of goal maintenance, performance monitoring and response inhibition.

Overall, many studies lead us to believe that the prefrontal cortex is involved in creating and maintaining goal representations. We can also reliably say that different parts of the PFC do different things within this area. Goals can be both extrinsic (typically, given to us) and intrinsic. The difference is paralleled by a neural distinction of lateral and medial prefrontal cortex networks, as seen in Lieberman's work (Lieberman *et al*, 2007). In 2006 Johnson led a study that showed greater medial prefrontal cortical activation while the subjects reflected upon their personal hopes and aspirations rather than distraction (Johnson *et al*, 2006). This has links to internal goals conceptually. On the other hand, extrinsic goals are linked to the lateral PFC.

The classic Libet experiment, which showed that a readiness potential in the brain occurred 500–1,000 milliseconds prior to the conscious experience of intending movement – has links to goal-realizing behaviour. This is almost like our brain knowing that we intend to do something before we do. It has raised a lot of dialogue around the notion of free will. Inhibiting responses is vital to the achievement of goals. For example, stopping ourselves sitting down on the sofa before going for that run, or stopping ourselves snapping at that colleague before we have considered a calm response. Conscious inhibition relies on the prefrontal cortex.

A study by Spunt and Lieberman (2014) highlighted some important things to be aware of practically with goal setting. They recorded the brain activity of subjects who were watching videos of people doing things such as brushing their teeth or reading a newspaper. The subjects were told to think about how the people were doing the tasks and then why they were doing them. The 'how' thoughts corresponded with areas on the left side of the brain involved in planning movement and tracking location (the premotor cortex and posterior parietal cortex). When they considered 'why' then areas involved in thinking about states and intentions of others lit up (the right temporoparietal junction, the precuneus and dorsomedial prefrontal cortex). Considering different elements of a goal activates different areas of the brain. We look at what you as a coach should do about this in the next section of this chapter.

The study 'Effects of three core goal pursuit processes on brain control systems' by Berkman, Falk and Lieberman (2012) identified functional activations distinct to each of the processes. They also reported interactive effects between response inhibition and goal maintenance in the dorsal anterior cingulate cortex and between performance monitoring and goal maintenance in the superior frontal gyrus and supramarginal gyrus.

When we focus in on our ability to maintain a representation of the goal in our working memory so that we can take action linked to that goal, we are talking about goal maintenance. The Stroop task, which normally activates the dlPFC and ACC, gives us some more interesting information. The Stroop task involves people naming either the word they see or the colour of the word. This test is actually requiring participants to do both goal maintenance and response inhibition (when the word and colour of the word cease to be the same). For example, the word red will be shown in red, the word yellow in yellow, then perhaps the word red will be shown but in blue. What we see is that the dlPFC is only active when doing the first part of the test, where goal maintenance is occurring (not response inhibition).

When we focus in on performance monitoring we know that we are asking the brain to identify gaps between where we are and where we want to be. It requires us to adapt and proactively get back on course. The dACC was shown to be involved in both goal discrepancy detection and reduction. More recently the dACC has been shown to be more active when it responds to an incongruence in a sea of congruencies (eg the word red appearing in the colour blue). Now it is understood that the ACC activation signals the discrepancy but tells other brain areas to reduce that discrepancy – almost like a head bouncer who sees conflict brewing and then gets on his walkie-talkie to tell other bouncers to go in and sort it out.

When focusing on response inhibition, evidence points us in the direction of the right inferior frontal gyrus (rIFG). With studies that involve inhibitory tasks (eg task switching, go/no-go stop signal) it is thought that the rIFG is the only brain area necessary for inhibition. However, several other areas are also activated, including the dlPFC, vPFC and ACC. Newer findings suggest that the anterior insula and parts of the striatum (head of caudate) are involved in inhibition.

Implementations intention

Frank Wieber and Peter Gollwitzer propose 'implementation intentions as an easily applicable planning strategy that can help overcome procrastination by automating action control' (Gollwitzer and Wieber, 2010). In other words, implementation intentions are a valuable tool to support clients who are working towards goals. They are important in specifying the details that will underpin the achievement of a goal. Let's take a seemingly simple goal of flossing your teeth. Many people share the goal of flossing their teeth regularly; however, far fewer do it as often as they would like. An implementation intention can help, especially one in an 'if-then' format. Consider the benefits of programming yourself to pick up the floss every time you put toothpaste on your toothbrush. So 'if I put toothpaste on my toothbrush, then I put it down and pick up my floss and use it'. What about when you feel in a hurry and are tempted to skip flossing to save time: 'If I'm in a hurry then I floss while thinking about the time I'll save from not getting tooth ache and having long-drawn-out dental treatments.' The great thing about 'if-then' intentions are that you can create them for many eventualities, and indeed considering these eventualities will stand you in better stead for dealing with them.

Implementation intentions come with a wide body of research that shows their benefits. Studies have shown that they have effects over and above motivation to succeed. This was demonstrated when participants who formed temptation-inhibiting implementation intentions outperformed the groups who didn't have these in place. Other studies have shown that they can help overcome self-regulation depletion (or willpower burnout). Participants of one study had to suppress their laughter during a funny film; afterwards they were supposed to do a task involving self-regulation (solving a series of anagrams). Normally you find that after people have been exercising their willpower they get depleted and find such tasks to be more difficult, so they would solve fewer anagrams. When the participants are given an 'if-then' implementation intention, such as 'if I solve an anagram, then I will immediately start to work on the next one', then they were able to solve as many anagrams as the group who hadn't been depleted.

Process versus outcome mental simulations

Most coaches are experientially familiar with the impact of thoughts on action. One study looked at the difference between mental simulations that focused on the process required to achieve a specific goal versus the outcome of the goal itself. This study was done with college freshers for five to seven days prior to an exam. They were split into three groups and mentally simulated, by either having good study habits or getting a good grade or both. Results showed that by focusing on the process the students got better grades; the way they did that – it is suggested – was by enhanced planning, studying and reduced anxiety.

Older studies have shown that by imagining events (rather than just reading about them, for example) people have greater confidence that they will occur. We know that the brain and body respond when we imagine things.

Having an emotional response to the mental simulation of the process required to achieve a goal, or the goal itself, seems to be a positive thing. It can serve to motivate clients, give valuable feedback and lead to improved action.

Willpower

The neuroscience of goals would not be complete without looking at the topic of willpower. We cover this in depth in Chapter 25, but mention it again here because it is so important to the current discussion on goals. Here we will just reinforce that a person's ability to take the small steps that will lead to the achievement of a goal can be the difference between success and failure. Perhaps an initial goal for any coaching client should be to improve their self-control! A study that looked at 92 adults and their capacity for self-control showed that it is possible to improve this by regularly practising small acts of self-control. This study involved the participants cutting back on sweets, or squeezing a handgrip, things that required their self-control. Their self-control was assessed before and after a two-week period using the stop signal paradigm. Something as simple as switching to clean your teeth with your other hand can help improve your self-control.

How is it useful to managers who coach?

Managers have the opportunity to guide their teams' mental processes, often on a daily basis. The power of our brains is immense, and managers can help their people tap into this power. Research tells us that imagining the process, rather than just the result, is most effective in helping us achieve that result. Managers can use this in meetings. When a shared goal is set, talk people through the steps to get there. Better still, ask different team members to paint the picture of each step. When you have one-ones with people ask them to describe the process they will go through to achieve a KPI.

As a team could you use implementation intentions? It could reduce unnecessary 'in the moment' decision making and help everyone move towards more desired behaviours. If wellbeing is an organization-wide goal (as it seems to be more than ever for many organizations) then how can you work collectively to prioritize this? Could it be that midway through any team or zoom meetings you have a 5-minute break where everyone is encouraged to go out into the fresh air. This might be the only 5 minutes some people take in a day.

What can I do with a client or colleague this afternoon?

- Setting goals that are widely unrealistic can motivate some people but unmet goals can be very demotivating to other people. So, initially, the safe bet is to support clients in setting realistic goals so that they get some wins under their belt. Offer clients the opportunity to increase their self-control by working on it for a two-week period, explaining that this should make it easier for them to achieve their goals.

- Within a full goal you need to know why you want to achieve it, what you are seeking to achieve and how it will be achieved. These should be separate components, as the brain struggles to focus on them all at the same time.

- It can be useful to move up and down within the goal hierarchy from 'why' to 'how', in order to get unstuck from a task or to become more practical by focusing on the 'how'.

- Another thing we can take from the research into goal maintenance, performance monitoring and response inhibition is that it could be useful to keep goals in mind externally rather than relying on working memory. So a smoker who wants to quit might find that rather than having to remember this when their cravings strike, instead to have a visual cue somewhere, such as a screensaver on a phone or computer. Perhaps a picture or comment placed actually on a cigarette box may help. Similarly for weight-loss goals, consider having a picture of either the ideal figure or part of the process (such as eating healthy snacks) on the fridge door.

For a goal like 'go to the gym regularly' there are several things a person can do to make it easier to go. One could:

- Go to the gym and do everything you can to enjoy it (play your favourite music and tell yourself how great this is making you feel).

- Connect to the sensations of stronger muscles while doing weight repetitions, feeling your legs moving quicker on the treadmill, the tightness of your abdominal muscles after your crunches (this connects you to the benefits each and every time you go, rather than just in three months' time).

- Practise thinking about the gym while getting yourself into an excited and interested state.

- Anchor the state and the thought together by listening to an upbeat song or spraying a special gym deodorant (that you keep only for using just before you drive to the gym).

This topic has links to:

- optimism;
- willpower;
- dopamine;
- neuroplasticity;
- priming;
- nudging;
- habits.

Christian on goals

I asked Christian what he thought of goals in coaching.

I've got two answers to that, depending on the audience we're talking to. There was a time when I would define coaching as a process of setting and achieving goals, and in some cases it is that. But I'm increasingly thinking goals may be unhelpful in coaching and get in the way. I can think of a particular client who I worked with over a six-month period. The goal was very clear from the outset and we pursued it diligently and after six months my client achieved their goal. However, it was a topic that I took to my coaching supervisor because I was feeling a little bit uncomfortable. Although they had achieved their primary goal, there were costs to achieving it. I had helped with a single-minded, complete focus on achieving the goal but maybe at the expense of other relationships, other sub-goals etc. I'm still curious about that because I didn't know whether it was right. I think my client was very satisfied – that's the feedback I had – so it's interesting that as a coach I had some doubts that I took to my supervisor – I'm not sure if my coachee had similar doubts.

I've been thinking more deeply about this because I've finished writing a book with a colleague called *Coaching in Islamic Culture*. Our new methodology is called Ershad. One of the things we've removed from the coaching process completely is goals, which probably then made me reflect on the kind of coaching I've been doing. For complicated reasons, in Islamic culture it was felt inappropriate for coachees to set a goal for themselves without any other factor being taken into account, so we thought a lot about what would replace goals and that goals are particularly limiting anyway. A goal is quite challenging nowadays with the VUCA world we're in. There is actually a risk that as you're pursuing the goal the context changes and therefore having a clear goal and sticking with it, in some cases, could be counterproductive and lead to more stress and tension. In the Islamic work we've replaced the idea of a specific goal with an intention. I prefer 'intention' to 'goal' because an intention is a broader kind of view. You can almost call it a strategic direction. It's a direction of travel. I'm also using more and more the concept of the desired outcome, getting a sense of 'That's what I'd like to be moving closer to.' It allows for things to change – and they are going to definitely change – and for slight corrections of course as people are working towards it. Another really important point that's come to my coaching more recently is the idea of meaning and purpose. If you talk about intention it's possible to link one's intention with meaning and purpose, and therefore it's almost possible to do a little check-in on 'Is this intention aligned with who you are and what you want to do?' That's really powerful, so nowadays when I talk about goals I might use the word intention: 'What's your intention out of this coaching?' When they explain to me what the intention is I can say, 'How does this fit in with your values or your purpose?' For me the alignment between somebody's intentions and their values or purpose or principles is important for well-being, and would allow coaching to then be about the pursuit of objectives, intention, etc, but without losing sight of one's values, principles and who they are.

Habits 35

What is it?

Habits are regular tendencies or practices. They are often unconscious and automatic. We each have lots of them, some useful, some not so.

Why is it important for coaches/managers?

Habits are naturally great things that free your client's mind to focus on new things, making them both more efficient and productive. They enable us to cope with the huge amount of information, choices, decisions, goals and emotions we experience every day. We have habits for a large part of our daily life. Many habits are unconscious, which means they can be directing people's behaviour without them being consciously aware of this. While good habits make it easier to achieve desired results, because we naturally do these things without conscious thought and less energy is required to fall on these defaults, undesirable habits make it difficult to achieve results.

What is the underpinning neuroscience?

One of the important things to be aware of here is called Hebbian theory, which underpins Hebbian learning. A Canadian neuropsychologist called Donald Hebb theorized that:

> When an axon of cell A is near enough to excite cell B and repeatedly or persistently takes part in firing it, some growth process or metabolic change takes place in one or both cells such that A's efficiency, as one of the cells firing B, is increased.
>
> (Hebb, 2002)

This has been simplified to 'Neurons that fire together, wire together'. As you have repetitive thoughts or take actions repeatedly your neurons fire again and again, forming a stronger and stronger neuronal circuit.

The formation of these circuits has the benefit of meaning we can do things on autopilot. The downside is that many of the circuits that are created are undesirable to your clients. The more that a habit is performed the stronger it becomes. This is because when the neurons fire they attract a protein blend called myelin to insulate it. This speeds up the transmission of signals, in effect making it easier and quicker for the habit to occur next time.

The neural pathways that encode habits are found in the basal ganglia, so this is the part of the brain often associated with habits. However, recently it was discovered that habits are not exclusively automatic. A study from MIT showed that a small part of the prefrontal cortex, called the infralimbic cortex, is involved in shutting off habitual behaviour (Smith *et al*, 2012).

What are any interesting studies in this area?

One area of research linked to habits is that of people with addictions. It has been claimed that people with gambling addictions, for example, have habits that they have no control over. Reza Habib, a cognitive neuroscientist, did some research that increased our understanding in this area (Habib *et al*, 2010). The experiment involved some serious gamblers whose lives were being affected by this habit. These people had lied to their families and had missed work in order to gamble. The other people in the study gambled socially but did not have any problems associated with this habit. Each person went into an fMRI scanner and watched the wheels of a gambling machine turn around on a video screen. This slot machine would either deliver a win, a loss or a near miss. There was no money involved in this experiment; the idea was that the gamblers simply watched the screen while their brain activity was recorded.

What they found was that the brains of the pathological gamblers got more excited about winning. The people in the experiment knew they weren't actually winning any money in this instance, but their brains still responded similarly. The areas related to emotion and rewards were more activated in the pathological gamblers than in the social gamblers. What the researchers found even more interesting was what happened in the brains of the two types of gamblers when there was a near miss on the screen. The pathological gamblers' brains registered the near misses very closely to how they registered a win. The social gambler registered the near miss similarly as a loss.

As we know as coaches, different people can view a single event very differently. This intricacy of the brain makes managers' jobs more difficult. There is corresponding neurological activity that underpins these differences. What may on the surface seem simple can be more complex when we cross-reference with what the brain is doing. The responses of our brains can trigger habits. So when a gambler wins and gets a mental high then they are triggered to play again to get another high. For pathological gamblers, the same is unfortunately true when they get a near win.

A key thing for consideration when working with habits is the preparation for what happens when an old habit is reverted to. If people don't know how habits work then there is a tendency to feel that they have failed, to feel worried they'll never be able to stick to the new habit or simply to feel guilty if an old habit creeps in.

Ivan Pavlov did some interesting research that gives us an insight into why habits sometimes return. In one set of experiments he found that if a rat was conditioned with a tone and electric shock in one box, and then moved into another box, the fear response could be 'removed'. If the rat was put back into the original box then when it hears the tone again it would experience the fear response again. It was also found that stress could reinstate a previously extinguished response.

Basically, habits tend to be deeply ingrained. When you stop performing a habit, such as snapping at a colleague whenever they interrupt you, then this neural circuit gets weaker. However, we don't believe it disappears completely. When you create a new habit and strengthen it over time and with lots of repetition, for example responding to any interruptions with a hand indication and the request for the person to 'give me a minute', then under normal circumstances this becomes the natural response. However, when under stress it is possible that the old circuit becomes favoured again and a return to old behaviour is experienced.

Research from the National Institute on Alcohol Abuse and Alcoholism offers some additional insights into breaking habits (Crews *et al*, 2009). This ability is hugely valuable to people both at work and everywhere else in life. By being able to create and enact new habits rather than default to old ones individuals can access flexible decision making and can align more effectively with goals. You might remember that other research has shown that the dorsal medial striatum is involved in goal-directed actions and the dorsal lateral striatum is needed in habitual actions. This particular study found that the orbital frontal cortex is critical in shifting from habitual behaviours to goal-directed actions. This is an area of the brain that is involved in obsessive-compulsive disorder, where unhelpful habitual actions are adhered to compulsively. This study was performed using mice, and the process called optogenetics, enabling scientists to activate individual neurons.

An experiment conducted at MIT looked at this area of breaking habits, this time involving rats. The study first conditioned rats to run a maze in a specific way. They checked that the habit of running the maze in this specific way had been thoroughly ingrained before moving on to the next stage. Normally we think of habits being linked to the basal ganglia, which they are. There is also involvement of a small area of the prefrontal cortex known as the infralimbic cortex. Using this advanced technique again, optogenetics, the researchers were able to inhibit specific neurons using light in the infralimbic cortex for seconds at a point where the rats were deciding which path to take next. The effect was that the rats stopped their habit and chose to run a different way. One researcher said that by turning off the infralimbic cortex it switches the rats' brains from an 'automatic, reflexive mode to a mode that's more cognitive or engaged in the goal'. This is good news for coaches because it means that there is a degree of executive function involved in breaking habits.

When the rats formed a new habit the researchers found they could switch this one off too by again inhibiting the infralimbic cortex. The rats reverted to the previous habit. This is evidence that old habits are not forgotten, there is a representation of the habit still intact within the brain and this can be retrieved instantaneously.

How is it useful to managers who coach?

Habits are your friends. At least you want them to be. When harnessed effectively, they can unlock huge potential in your people. Reducing cognitive load and making advanced decisions can reduce pressure on the PFC and unlock more of a High-Performing Neural Environment. As a manager you can consider what organizational and team level habits will serve your people. How can you help them free up their cognitive reserves to use on innovation, quality thinking, solving problems and so on?

What can I do with a client or colleague this afternoon?

Due to the nature of habits there are lots of things that you can work on with your clients and colleagues. There are also lots of links to other concepts within coaching. A great approach, depending on your client and the situation, could be to focus on a specific habit. In order to decide where to focus you could:

1 Invite your client to verbally share with you five habits that they feel are holding them back or doing them a disservice in some way.

2 Once you feel that you understand these habits, the triggers for them and the negative effect they are having, you could then move on to get clear on what habits would be supportive in reaching desired goals. Perhaps elicit between 3 and 10 here. You could ask your client to write these down either during a session or, even better, at home. Again, you want to know the value that these habits will bring to the individual. The next stage is like being an architect: the aim is to identify all the small parts to one of the new desirable habits.

3 Identify small steps involved in creating a new habit.

Then, depending on how successful your client normally is with establishing new habits, you want them to start implementing and mastering one step at a time. So this stage can be really slowed down for people who might have a history of not creating solid new habits – literally taking one component and working on it until it is second nature before introducing another step. Another point to mention here is that this process is best initiated when life is not overly stressful. If your client is trying to move house, lose weight and start coaching their team at work in a new way then it would be wise to wait until the house move is complete and the weight lost.

There are a few other things you could do with someone you are working with that could be useful:

● Help them understand how a habit is formed and therefore how to avoid giving attention to undesirable habits.

● Give comparative frames of reference of people who have rewired their brains: for example, people who have regained the use of paralysed limbs (this helps expand what your client thinks is possible). Check out **www. neuroscienceforcoaches.com** for more ideas and resources to use with clients.

● Explain the neuroplastic nature of the brain so that they are clear that habits can be changed.

PART SEVEN
Neuroscience of not-so-classic coaching areas

Part Seven looks at some tools and concepts that are not traditionally used by coaches or necessarily consciously addressed by coaches. Something like *trust* is, of course, very important in every coaching relationship but is perhaps not paid attention to in other relationships unless there is a problem. Here we get to see some of the far-reaching effects that trusting and being trusted can have on people. The opportunity then becomes one of examining whether actions to increase trust or trustworthiness would benefit the individual.

In this part of the book it is even more important to remember that we are focusing on the definition of coaching as facilitating 'self-directed neuroplasticity'. Like *trust*, things such as *fairness* or *loneliness* do not tend to be focused on unless there is a problem. Coaching is often employed to improve things rather than just as a remedial solution, therefore there is a clear chance here to really add to a coach's toolbox.

The areas of nudging and priming have their roots in psychology. They are very useful for coaches to be aware of to help effect behaviour change without needing to rely on willpower. Leaders and managers who are aware of these techniques have a real advantage.

Christian: What else should we be considering?

I asked Christian: 'Are there other conditions that you think are critical for good coaching interaction?'

There are some necessary conditions. Not conditions for change or for coaching, but conditions for the right kind of relationship that will allow people to think deeply and carefully about important decisions they have to make. One of the conditions is trust. I think there needs to be mutual trust between both parties. I guess from the coachee's point of view the important thing is to trust the coach and their ability to stick to the contract, to keep confidentiality, to know what they're doing and to be appropriately trained, but the coach also has to trust a couple of things.

One is, I think the coach has to trust that the coachee is telling the truth, and that comes up a lot in conversations. A lot of trainee coaches say, 'What if the coachee isn't telling the truth?' And that has never occurred to me when I'm sitting there talking to a coachee who paid to be there, who is struggling with something or dreaming about something – I trust them. I trust they're bringing me what's at the forefront of their mind but I know that may change. I think there also needs to be mutual respect between the coach and the coachee. The mutuality is important because if the coachee has some doubts about the coach's professional credibility, I think that gets in the way and has to be resolved one way or the other. I also think it's even more important that the coach is respectful of the coachee, their culture, the situation they're in, and respectful of the challenge they're facing or the ambition they're going for and one of the parts of that respect is not believing that the coach has the answers. It's possible for me to suddenly come up with a solution for something that's very complex but part of being respectful is not offering advice because, for me, the contract of coaching is not advising, it's coming back to that idea of helping people to learn rather than teaching them. So I think those are two things that are essential for me, trust and respect.

This is a really interesting point Christian makes about coaching not involving advising people. From working with lots of coaches I think opinion is divided. Robert Dilts, author of *From Coach to Awakener*, suggests coaches put on many different hats during the coaching journey and at times teacher may be one of them. There will be a fine line at times, as our questions can of course be leading.

I asked Christian what qualities he thinks make a good coach, as I noticed he mentioned being humble before.

The coachee has to respect the coach's professional credibility, which almost challenges the idea of humility. I do think of course the coach must be humble and must enter into the conversation from the position of humility and of 'I don't know the answers'. I think the coach should always be contemplative. When you're coaching someone it's possible for little thoughts to pop into your head, some sort of analysis of what you think might be going on. Or things might pop into my head like it might be interesting to use this particular tool or particular set of questions. I think even then the humility is thinking: Well, I might be wrong and being very contemplative in all of those things. Humility allows you to be open and curious and my sense is that it seems to translate well across cultures as well, as people can pick up when somebody seems to be quite humble and open. Humility is definitely a quality that I think a coach should have.

They should have curiosity as well, but there is a balance. I shouldn't be asking questions of my client because I'm curious about everything: 'Why did that happen? Why don't you do this?' It's not that kind of curiosity that leads to enquiry. I guess the most important curiosity for the coach is: 'I wonder how this person is going to find a way to get closer to their desired outcome?' That kind of curiosity is important, with the coach always thinking, 'They've suggested this. I wonder if it's going to be the right thing?' or 'I wonder if maybe…' Integrity is important too. Many coaches nowadays have portfolio careers so they will be executive coaches but they may also be consultants, trainers, and a number of professional roles and I think sometimes those roles can blend into one another, so the coach has to have integrity so that the coachee trusts them. All the decisions made should be in the best interest of the client and integrity means not encouraging things or pursuing particular paths that are going to be better for the coach, if you know what I mean.

I asked Christian: 'In your opinion what makes coaching powerful? You used the word "powerful" previously.'

I do think it's powerful. When I go and work in organizations I come along and say, 'I think coaching would be quite a good idea.' And then some of the pushback is, wouldn't it be better to train them and the answer is yes, if what you want is transfer of skill, you should train them. And then they'll say isn't it better to mentor them? And again I'd say yes, if the idea is to transfer knowledge from some experienced person to a less experienced person that is the right way. In which case the question is why would we use coaching? And I think coaching is powerful because it empowers coachees and it might be one of the few contexts in which leaders and other influential people actually withhold their knowledge and information and advice so that it creates a sense of self-reliance, self-confidence and self-esteem. Organizations really have to be ready for what coaching may bring because if an organization has a command-and-control structure – very hierarchical, very rigid – then coaching is not the right approach. Coaching is all about helping people to come up with their own ideas. It leads to innovation and people asking more questions and challenging the status quo, so it is powerful in that sense and it can bring change. Organizations at least need to be open and ready for change or even encouraging it, as coaching is a great way to encourage people to take control.

Some of Christian's current research involves the environment people are actually being coached in. I asked Christian what prompted him to start exploring this.

I saw Ilona Boniwell give a keynote at a conference and one of the things she was talking about was the use of play in coaching. She was talking about Lego's Serious Play. Her argument is that we learn a lot about each other when we play so if you can sit in a room and play with people around the kind of changes they want it can be really powerful. So I've been doing a lot of reading and a lot of my own thinking about what makes coaching effective, and in *An Introduction to Coaching Skills* I argue you need three elements together, which can lead to effective coaching. One element is the set of skills and other things that are trained on regular coaching courses. Another is a process, and when I came up with this initial theory I was thinking narrowly of the GROW process or something similar. The third element is a way of being, which is what we've been talking about: being humble and considering how the coach should be as a human being. I think that's really powerful and important but for me the process now seems to be more than just what happens during the conversational process; we should think about what needs to be happening during the conversation.

A lot of modern executive coaching can probably find its roots in counselling and maybe even in psychotherapy. Another root is probably sports coaching which is very different again. I've been wondering where the concept we have came from, that the best way of coaching is for two human beings to sit in a room, with two chairs and a table or not even a table. Where did we get the assumption that this is the best form and best kind of context for these conversations to happen? I'm speculating, but maybe it comes from counselling and then psychotherapy before that, where this did seem to be the traditional way of having these conversations. That liberated me a little bit to think, well, modern-day coaching isn't counselling and it doesn't have to follow that pattern. My own experience also raised questions about that because sometimes I must do coaching in large organizations, commissioned by the company, where I go and meet individuals. I experimented with a lot of things and the least productive for me was going to their office, as there were interruptions and no sense of equality. So I experimented with different spaces.

I've experimented with going into a hotel lobby or going into a coffee shop or going into a meeting room and I've noticed that actually where we are does make a difference in my own experience. Often in a work context the challenge or the issue is something that actually happened at work and most of the time my clients tried to resolve that at work, usually sitting down at the desk, so I thought maybe a different environment would be helpful. To some extent going to a coffee shop was helping, but we were still sitting and talking. When I was overseas once my coachee and I went for a walk in some botanical gardens. That was partially because I wanted a bit of fresh air but so did my coachee and that worked really well and I've carried on trying that now and then. That made me think that the process bit is not just a conversational process, but could include things like going outside, walking on tightropes or swimming. I've been thinking about what the best context might be for people to think more creatively or differently, because usually that's what they are there to do. As a result of that we are actually in the middle of a research process and we've just got ethical approval. It's a qualitative study but I'm curious about the experience of people who aren't normally coached when sitting. We're going to get pairs of coach and coachee to come to Henley Business School, and instead of having a seated conversation we're going to ask them to coach while walking. What I'm interested in is people who are used to being coached while seated and I want to get a sense from their own lived experience of what they feel is different when the same coach they always have coaches them in the outdoor space while walking. I think walking is an important part of this and I'm interested to see whether there's some themes that come up from the experience of a number of coachees who experienced this and then if there are themes then I think it's worth exploring further.

Choice architecture 36

What is it?

Choice architecture is the process of affecting outcomes by influencing decisions. Typically this is done by presenting choices in a specific way or by making changes to the environment within which someone exists. There are many tools in a choice architect belt; we will look at just two here. We run Behaviour Change training for organizations both as a full course through Synaptic Potential and as a module in a leadership programme with Cambridge's Moller Institute. If this piques your interest, you'll find a white paper on this topic at **www.synapticpotential.com**.

Nudging has been described as getting people to make decisions based on how their options are presented. According to the iNudgeyou network:

> The following conditions are necessary (and perhaps sufficient?) of X to be a nudge within a certain choice architecture:
>
> **1** X alters the decision maker's behaviour in a predictable way
>
> **2** X does not limit the choice set, ie the available options to the decision maker
>
> **3** X does not significantly alter the decision maker's incentives, in terms of payoffs and utility increase from the perspective of a rational agent.

X can be a huge range of things. Singapore is a country where nudges are all around you. A beautiful example is where stairs in the Bugis MRT station are brightly coloured. They also say various things like 'I want to climb the stairs to fitness.' There are escalators right next to the stairs, but when they are not being used they are still and quiet. The result is that many people choose to take the stairs.

Nudging finds its home within the field of behavioural economics. This is where the fascinating realization that human decision making is not a purely rational process is explored:

Nudges are ways of influencing choice without limiting the choice set or making alternatives appreciably more costly in terms of time, trouble, social sanctions, and so forth. They are called for because of flaws in individual decision-making, and they work by making use of those flaws.

(Hausman and Welch, 2009: 126)

Nudging is a big and exciting field. There are specialists experienced in designing nudges, gathering and evaluating data and redesigning nudges. For leaders and managers of organizations who are looking to implement wide-scale nudges that have a lot riding on the results it would be wise to consult the experts. For smaller-scale interventions as a coach you may choose to experiment.

Priming is the second tool we will look at. We can think of the process as activating certain neural circuits and setting us up to respond in a certain way. Technically it is an implicit memory effect where a stimulus influences you.

Why is it important for coaches/managers?

Nudging is important in several ways to the people you work with. First, as a coach you can suggest things to them, or ask them questions to direct their attention to doing something that will serve as a nudge to them. For example, if you were talking about going to the gym after work, a simple nudge would be to change into exercise clothes at work before getting into the car. This would not limit the individual's choices, they could still drive home if they wanted, but it is likely to result in more trips to the gym. You wouldn't need to explain all the background to a client every time you suggest a nudge. You could even ask questions to help them come up with a nudge, depending on your chosen coaching style.

Second, you could filter for ways that those you work with have unhelpful nudges set up. A leader or manager in a coaching role may see or hear things that are not creating the best environment for individuals. Perhaps people want to be fit and healthy, which is obviously great for productivity, low sick rates, etc. Maybe the layout of the building is such that most people have to walk a long way to take the stairs between floors whereas lifts are right there and perceived as quick. Remodelling the office may not be a viable option, but putting up a sign that says something like 'Treat yourself to thinking time, take the stairs' might result in more people doing so. The exciting thing with nudges is that you can trial them and tweak them.

Alternative ways that you can use nudging is to support those you work with in achieving their goals. So would a motivational e-mail once a week or sporadically help your client? Would a reminder – one that they can print out and put in their work space – help them?

Priming is occurring unconsciously many times during each of your client's days (and yours!). Becoming aware of it and both cushioning against unhelpful primes and utilizing potentially helpful ones is sensible. The power and effectiveness of priming could do with more research, so this is not yet set in stone from a scientific perspective.

What is the underpinning neuroscience?

Nudging involves bypassing the cognitive system; it sidesteps any issues with reactance and attitude. You don't have to persuade anyone to do anything, simply communicate directly with the systems that control their behaviour. As coaches and managers we are extremely familiar with people's difficulties in changing their habits. The desire to change a behaviour is often met with challenge when the old neural systems that control the habitual behaviour continue to be 'switched on'.

In essence we can think of there being two systems that process different types of information for different aspects of behavioural control. One is the effortful and intentional, while the other is implicit and routine. Evolutionarily this enables us to perform routine behaviours unconsciously, thus freeing up our cognitive system to process other things.

In recent times there have been a lot of questions around priming. Several studies that were done have not been replicated, which casts doubt upon their credibility. The critics have suggested that there has been publication bias, the experimenter effect and that criticisms have not been dealt with well.

There is a little neuroscience that can help us out, although not much yet. In general it makes sense that activating a neural circuit in some way using a prime could make it easier for a linked circuit to be activated shortly afterwards. A study showed that conceptual priming is linked to reduced blood flow in the left prefrontal cortex, which is thought to be involved in the semantic processing of words. Another study suggested that perceptual priming is controlled by the cerebral cortex (specifically the extrastriate cortex). This study showed that perception is improved by prior experience of a stimulus. There was a reduction of gamma frequency oscillations... the assumption is that there are two stages to a prime's effectiveness in the visual system: 1) representation sharpening in the early visual areas; 2) competitive interaction between representations in the higher visual areas and the pre-frontal cortex.

What are any interesting studies in this area?

Many of the nudging studies do not have obvious applications to coaching; rather it is the process of nudging itself that you can then teach or apply imaginatively with clients and colleagues.

A classic experiment involved placing a sticker of a fly in urinals in Schiphol Airport. This Amsterdam airport managed to reduce the spillage, due to bad aiming, by 80 per cent. The UK government had taken notice of this field and had formed a 'Behavioural Insights Team' dedicated to implementing ways to nudge people (that has now become a private entity). Their work resulted in changes to forms and processes in the area of tax collection. This resulted in a 15 per cent boost to tax repayment rates. They also changed the UK driving licence forms to make it 'required-choice' for people to take a stance on organ donation (opting either in or out). This doubled the number of people who are joining the register.

A great series of studies in 2005 by Randy Garner looked at how to increase response rates of correspondence. One experiment looked at what effect it would have if a Post-it note was attached to a survey that was sent to faculty staff at a university. Their aim was to get more staff to fill in and return the survey. They did indeed find that by putting a hand-written request on a Post-it note for individuals to complete the survey, the response rate was significantly higher than when compared with either handwriting a request directly on to the front of the survey, or attaching a blank Post-it note (Garner, 2005). So the combination of Post-it note *and* hand-written request was the key here.

A favourite study is one that helps people to be more honest. Lisa Shu led a team that looked at what effect was had by getting people to sign at the top of a form declaring their honesty. Participants were told to report on their earnings from a problem-solving task and were also told that travel expenses would be reimbursed and they would be compensated for travel time. The results were stark. Of those who did not sign any declaration of honesty 79 per cent cheated. Of those who signed at the end of the form 63 per cent cheated. From the group of people who signed at the top of the form only 37 per cent made false claims. The reasoning behind this could be as simple as bringing ethics and one's own identity to people's attention before taking some form of action.

One final type of study shows the importance of placement of food. Several variations of this study exist now, and many more could be on their way soon. The study involved rearranging buffet tables to nudge people into eating more apples and fewer brownies. Figures in the range of 80 per cent increase in apples and 30 per cent decrease in brownies are not uncommon in this type of experiment. The subtle changes in our environment really do lead to different behaviours.

In a classic priming experiment John Bargh and his team asked people to assemble a four-word sentence from a selection of five words (Bargh *et al*, 1996). For one of the groups the words included ones associated with the elderly: for example, Florida, forgetful, bald, wrinkle or grey. The other group's word selection didn't contain these words. After each individual had finished the task they had to walk down the corridor to another room to do another experiment. The actual experiment was them walking down the corridor: the time it took each of them to do the walk was measured (covertly) and what they found was that those primed with words associated with the elderly walked slower! This has become known as the 'Florida effect'. When the students were later questioned about their experience of the first task, none of them reported noticing a common theme among the words and all insisted that they couldn't have been influenced by the words. They weren't consciously aware of acting more like an old person!

This ideomotor effect works both ways: so walking slowly makes you more likely to recognize words linked to old age. A common task to lift your spirits is to hold a pencil between your teeth, parallel to your face, forcing you to smile. When students who were doing this were asked to rate cartoons, those who had the pencil this way round rated them as funnier. Other students, who had the pencil held at a 90-degree angle, forcing them to frown, found them less funny.

One of my favourite priming experiments that I cover in *Make Your Brain Work* is what I call 'How to win a game of trivial pursuits.' Two scientists did an experiment to investigate a link between stereotypes priming on overt behaviour. This involved priming people with the stereotype of a professor, a secretary and no priming at all. In this case the priming was done by getting people to read something and there were either strong mentions of a professor or a secretary or neither. The idea was that, stereotypically, professors are thought of as knowledgeable and intelligent whereas secretaries are not. The people were then asked to answer multiple-choice questions – from the game Trivial Pursuit.

The professor-primed people answered more questions correctly than the other two groups, and the secretary-primed people answered the questions most quickly.

How is it useful to managers who coach?

I often say to managers and leaders 'You are custodians of your people's attention'. I mean that they have power, responsibility and need to protect and direct attention wisely. There is huge opportunity in using the choice architecture tools from behavioural science to guide and influence team members' behaviour. People's brains are typically overloaded. Where possible, managers can reduce conscious decisions their people need to make and make it easier for them to do the desired thing without even thinking about it.

At Synaptic Potential, we use nudges frequently in learning and development programmes. Managers can think about the components of their ideal culture they want to invest in and design nudges around these too.

What can I do with a client or colleague this afternoon?

A great process is to sit down with a client and focus first on helping them identify the top three behaviours they would like to change. Look at what they are doing, when, why, how and make sure you understand what the triggers are to that happening. Elicit what the desired behaviour is instead. Explore what potential nudges could be trialled to unconsciously make the desired option more favourable. Either do this by yourself and make suggestions on what could be trialled for two weeks before evaluating the results, or brainstorm with your client openly. There is less research to show how effective nudges are when people are aware they are being nudged. This is worth bearing in mind but, as with everything here, experimentation is your friend!

Sometimes brainstorming with your client may result in a proposed nudge to help change other people's behaviour, which will then help your client achieve their goals. A classic example is: if a client is finding they are achieving less than they want to in a given day and can attribute this to frequent interruptions then they might choose to close the door to their office at certain times during the day. Very simple, yet often very effective.

On the priming front, challenge your client to open up their awareness for two hours a day for three days, and look at what comes into their consciousness. They won't notice everything (I'm not even sure that is possible!) but they may observe things like:

- When I watch the news, my mood seems lower afterwards.
- When I listen to upbeat music on my way home I work out harder on the treadmill.
- When I smile lots and talk calmly to my children they seem to be easier to look after.

Then ask your client to suggest anything they could put in place that may help them to be in certain states or act a certain way. For example:

- motivational words around their desk;
- walking briskly before a meeting;
- adding some specific pictures around their home and work areas.

There currently is not any direct evidence that this will work... but logically it shouldn't do any harm.

More on behaviour change from Marshall

Marshall Goldsmith, an American leadership coach, doesn't say that he underpins his work with neuroscience. However, neuroscience does support or predict many things he says. This makes total sense as he works with people with brains and neuroscience studies brains. So it all ties together that in Marshall's many years of experience he has seen things in organizations that neuroscience would predict from the lab.

The recognition of behaviours as critical to results is of vital importance. Many individuals and coaches may say they know that but act differently. Unless you are able to change behaviours you limit yourself. There are different approaches to behaviour change.

Marshall asserts why it is important for companies to commit to behavioural change in leadership: 'Number one, for you to become a more effective leader, and number two, it's important to be a role model. If you're the CEO of the company I'm assuming you want everyone else to try to improve. Well, the best thing you can do is let them watch you try to improve.'

Our scope is beyond CEOs but the points are still very relevant. A leader of a family (be that a parent or a child) is even more compelling if they are demonstrating effort to improve.

Marshall says that his work is largely about:

> the transition between being a winner and a high individual achiever, where it's all about me, to a leader, where it's all about other people: letting go of ego issues, letting go of having to be right, letting go of being a star, and trying to make the transition to being a leader who helps other people be the star. This is very difficult. The leader has to have courage to get feedback, which is not easy nor necessarily pleasant. They have to have humility to admit they can improve and the day-to-day discipline to follow what is required to actually change.

Many coaches work with high achievers. These individuals are frequently used to delivering value by contributing their ideas. The way their thoughts and behaviours have conditioned their brains is to release dopamine when they input. They feel good if they feel they are giving ideas or telling people what to do. This neural pathway needs a spin-off one that is strengthened. Leaders need to feel good about listening. They need to have dopamine fired when they encourage another person. That is also how they need to feel like they are contributing.

Marshall had an almost comical response when I asked what people struggle with most: 'Why people don't do what I teach.' Many people I'm sure can relate to not doing what they think they should.

> I did a study called 'Leadership is a Contact Sport' based on 86,000 people who have been to my courses. We measured whether they did what I taught them and not surprisingly the people who did the stuff I taught got better, and the people who didn't do the stuff stayed the same. The reason people don't do what I teach is largely summarized in a dream: 'I'm incredibly busy right now, given many pressures at work and home and technology that follows me everywhere and e-mails and voicemails and global competition. I feel about the busiest that I ever have. Sometimes I feel overcommitted. Every now and again my life feels a little out of control. But I'm working on some very unique and special challenges right now. I think the worst of this will be over in four or five months and after that I'll take two or three weeks and get organized and spend some time with the family and begin my new healthy life programme and everything will be different and it won't be crazy anymore.' I don't know if you ever had such a dream but that's very common among all the people I

work with. They're very nice people, incredibly busy, and getting them to do anything today is a real challenge. It is these delusions that can answer why we don't do what we know we should do. We say things like tomorrow's going to be different, high-probability, low-probability distractions. You never plan on low-probability distractions because, by definition, they're low on probability. Unfortunately, there are a million low-probability distractions that could occur.

As a coach you probably recognize this scenario. Recently I was teaching our Neuroscience for Coaches distance learning programme to a group in China. Someone said that they felt they had fallen behind as they weren't studying as much as they'd have liked. (It is completely optional what pace you choose to go at – if you do just two hours a week you'll finish easily within a year.) She was a lecturer in a university and there was always more they needed to be doing. Marking, researching, writing, seeing students, lecturing and so on. I made the suggestion that I've used helpfully for years which is to schedule in anything important – in a similar way you would something you'd never miss. For her, she wouldn't miss teaching a lecture and these are blocked out in her diary. We can use this. We can utilize the way her brain responds to the colour and format of that commitment. She can schedule a personal study session for herself each week in exactly the same way. While this advice has been circulating for many years (I'm sure I didn't come up with it) it is new to some people and poorly adopted by most.

Really simple strategies can put neuroscience insights into action. By understanding the brain we can be intentional and practical. I was speaking to another executive coach recently who typically works with leaders of countries. He told me that he often starts a session by inviting the leader to put all their meetings on to a long piece of paper. They can actually see how full their weeks are before anything else goes in. Many people continue to try to cram more things into a period of time than is possible. Then they say things like 'There just isn't enough time to do everything' and their cortisol levels rise steadily until they find themselves in a state of chronic stress.

Back to what Marshall thinks about adding too much:

A classic problem is that successful people, especially people with technical backgrounds – scientists, engineers, people like that – try to add too much value. So imagine I come to you with an idea and you think it's a great idea, but instead of saying it's a great idea you say, 'It's a nice idea. Why don't you add this to it?' The problem is the quality of the idea may go up by 5 per cent but the enthusiasm for the idea may go down 50 per cent. It's not my idea now, it's your idea. It's very difficult for smart, successful people not to try to add value. One of my coaching clients retired and he reflected to me about his leadership as a CEO and realized his suggestions became orders. If they're smart, they're orders; if they're stupid, they're orders – all his suggestions were orders. I asked what he learnt from me when I was his coach, and he said you taught me one lesson to help me be a better leader and have a happier life: stop and breathe and ask is it worth it? Half the time it wasn't.

My third book is called Engaged: *The neuroscience behind creating productive people in successful organizations.* It looks at a lot of practical ways people can create engaged brains. The way many business schools are still teaching people to run organizations doesn't take into consideration how brains work. From our experience and research at Synaptic Potential we believe that the best results are generated by taking an 'all bases' approach (we even train others to do this because it is so important). Working solely at an individual level or solely at a corporate level is never going to be as effective as covering both of these bases. There are many other bases to tackle – but as a coach you'll typically be focusing on the individual, which is why Marshall's next comments are interesting.

> If you look at the research, everything in employee engagement is focused on what the company can do to help you be engaged. I went to a presentation at the National Academy of Human Resources Conference and 100 per cent of the dialogue was about employee engagement as a responsibility of the company, 0 per cent was the responsibility of the individual – which is, to me, ridiculous. So it's like an individual has no control over their own lives. They're complete whiney victims of an over-oppressive system. My daughter examines questions. She's the one who got the idea of active questions and she said every question is a passive question. A passive question, for example, is 'Do you have clear goals? Do you have meaningful work?' Or my all-time favourite – 'Do you have a best friend at work?' The most ridiculous questions ever written. Do you have clear goals? No. Why not? They're confused. Do you have meaningful work? No. Why not? They make me do XYZ. Do you have a best friend at work? No, they're all jerks. But when you ask an active question rather than blame the environment you actually look at yourself and say Did I do my best? Our research on that has been amazing so far: thousands of people just asking active questions every day, and so they get better. Why? Because they focus on it.

> A lot of this is very parallel to Hindu philosophy. The whole essence of Hindu philosophy is you are here, you've been dealt a hand of cards and you can't change the hand, so your only choice is how you play the hand. Do your best. That's all you can do.

This mentality could certainly ease the burden for some people. A lot of the brain's energy can be used up thinking about things that can't be changed. If we let these things go, accepted them and felt empowered to go from there that is quite different. Acceptance and commitment therapy, training and coaching has some research underpinning why this is a beneficial approach (Arch, 2012).

While this all makes sense, the opportunity to tap into Marshall's experience was too good to miss. I asked him, 'What do you think distinguishes those who are able to let things go and make peace with them from those who struggle with that?'

> I think a lot of it is awareness, or you can use the word mindfulness. In life triggers constantly bombard us. A trigger produces an impulse and an impulse leads to behaviour. So that's one thing I do every day: I pay a woman to call me on the phone and she listens to me read questions I wrote and provide answers I wrote every day. She did it 10 minutes ago, this morning. Excuses don't help. It's hard. I pay a woman to call me on the phone every day because my name is Marshall Goldsmith and I'm the world's number one coach, and I'm too cowardly to do this by myself. I'm not disciplined. I need help!

I said: 'Amazing. That was very human of you. Do you see any big or small things that trigger people to take ownership of their lives? I'm hearing that self-awareness is important and I wonder whether there's anything you see which helps people take ownership?' Marshall replied:

> It sounds ridiculous but a lot of people have never thought of it. You are really surrounded by a culture that encourages victim hunt. The employee engagement industry is an industry that encourages victim hunt. Even the Gallup survey says if you're disengaged you must have a bad boss or work in a bad company. It's insane, it just tells people they have an excuse. If you want to act like a pain today it's not your fault.

What can organizations or individuals do to move forward along those lines?

> I'm not really an expert on organizations. I am an individual-level expert. So at the individual level what I teach is very simple: you get feedback, you pick what you want to improve, you take responsibility, you talk to people about what you want to learn, you develop a follow-up strategy, you measure improvement all the time and then, gradually, you get better. Not rocket science.

False memory   37

What is it?

We tend to think of memory as involving the encoding, storing and retrieval of information. There are several types of memory: sensory, short, long, declarative and procedural. While some of our experience of memory is conscious, most of what our memory is involved in is in the unconscious.

A false memory is a psychological term for a memory that might be a fabrication or distortion of an event – or a detail within an event – that didn't happen. It is also conceivable that emotions or feelings may be falsely recollected. Some factors that are known to influence false memory include giving misinformation, existing knowledge and other memories. It is possible to change existing memories to incorporate new information.

Why is it important for coaches/managers?

From a cognitive perspective, memories are involved in an awful lot. If memories are coded or recoded incorrectly this could be helping or hindering people.

People may strongly believe that if they remember something this means that it happened. Often problems arise when people act as if their memory is infallible. In a national survey in the United States people were asked if they agreed or disagreed with the following statement: 'Once you have experienced an event and formed a memory of it, that memory does not change.' Nearly 48 per cent of people agreed. It is quite widely believed that stressful events increase our ability to remember something. For example, someone tells their spouse that they want a divorce and a heated interaction ensues. Or at work someone gets a particularly bad performance review, which upsets them. It is not uncommon for people in high-stress situations to really believe that something happened when in fact it may not have done. Based on a lot of research we know that:

- Memories can be implanted.
- Powerful emotions can reinforce or weaken real memories.

- We can reduce the power of painful memories.
- False memories can elicit the same strong emotions as real memories.

The people you work with do not exist in isolation. Colleagues, friends, family members all interact and have the potential to influence their memories. The questions they ask, the summaries they give, the flippant comments they make can all shape what is remembered. You as a coach or a manager are also hugely powerful in your potential to affect how your clients remember things.

In a corporate environment, as in life, the stakes can be high. The research that tells us that the presentation of false information can influence what people recall in the future is very important. In a meeting, imagine that someone reports on the last month's events in his team. If he speaks of a conversation the group had last month and says he had mentioned some important data, perhaps believing he had shared this, it has an effect on the group. You are likely to then get division within the group. Maybe 40 per cent of the group will 'remember' him sharing this information, while the other 60 per cent will not.

Acting as if memories are 100 per cent reliable can lead to miscommunication and breakdowns in relationships. The consequences of this can be serious.

In an experiment done with children (whose parents had consented to the study) a man went to a playground and asked individual children to help him look for his lost dog (*Daybreak's Stranger Danger*, 2013). The children were shown a picture of a dog and a lead. Many of them walked out of the playground with this man. Later, some of the children when asked why they did it replied that the man had lost a dog. When pressed, several children said they had actually seen the dog. Interestingly with children, some people's first thought is that they are making up stories rather than they are experiencing a false memory.

What is the underpinning neuroscience?

The neuroscience of memories is quite complex. What follows are some of the highlights of what we think is happening. When you retrieve memory, information is being processed from two different regions of the brain: the medial temporal lobe (MTL) and the frontal parietal network (FPN). The MTL focuses on the specificities of the event while the FPN processes the overall gist of the event. A study using fMRI by Roberto Cabeza showed that when people recalled a true memory they had increased activity in their MTL, whereas when the memory wasn't true (but they still believed it to be) their FPN experienced increased activity.

What are any interesting studies in this area?

Elizabeth Loftus is the woman to watch in this field. She has carried out lots of studies into memory and is known as an expert in the field (see, for example, Loftus *et al*, 2002, 2013). One of her famous experiments is about implanting the memory of being lost in a shopping mall around the age of five years old. This was important because it was the implantation, the creation, of an entirely false memory rather than just a detail within a memory. The way they did it was to initially ask the study participants to try to remember childhood events. They created individual booklets for them with three one-paragraph stories about events that had happened to them and one story that hadn't. The false story was adapted or based on a real shopping trip that the individual had been on. Once the participant had read each story they were to write down what they could remember about the event, or write 'I do not remember this' if that was the case. Two follow-up interviews occurred next, where they were told that the level of detail they could remember, in comparison to a relative, was being examined: 68 per cent of the real events were remembered and 29 per cent of the false events were remembered. This study showed that it was possible, and relatively easy, to implant a false memory.

Another, almost comical, experiment was created in part to prove to critics that the events being falsely remembered hadn't actually happened (making it a retrieved memory instead). For this to work the thing that people were falsely remembering would have to be impossible. The researchers showed people an advertisement that described a visit to Disneyland and meeting and shaking hands with Bugs Bunny: 25–35 per cent of people claimed they had indeed met Bugs Bunny at Disneyland. Of these people 62 per cent remembered shaking his hand and 46 per cent remembered hugging him! Of course this is not possible because Bugs Bunny is a Warner Brothers character and not a Disney one (Loftus *et al*, 2002).

In a very different set of research, subjects were recruited from the US military who were enrolled in Survival School training. This is a programme that is designed to prepare individuals for the possibility of being captured as a prisoner of war. The training is stressful, including being hooded, spending time in a very uncomfortable cell and interrogation that is physically abusive. When they are rescued the soldiers are then questioned about who interrogated them and their experience, as would be similar to what would happen in real life.

During this interview process the soldiers were shown a photo of a person identified to them as the person conducting their interrogation. They were asked questions such as: 'Did your interrogator give you anything to eat? Did he give you a blanket?' In fact, the photo was not of the interrogator at all. Later, though, when asked to identify the interrogator 84 per cent of the soldiers identified the person whose photo they were shown. Similarly other information could be implanted into their memory: 27 per cent of them claimed to have seen a weapon that had not been there (after being fed misinformation); 90 per cent recalled a non-existent telephone. This phenomenon occurred in people who are highly trained to deal with stressful conditions.

By this point you may be wondering if anyone is immune from experiencing false memories. To date, researchers have not found anyone. A study at the University of California conducted an experiment with people who have a highly superior autobiographical memory (HSAM) (Patihis *et al*, 2013). This means that when quizzed these people could remember things like 'On what date did an Iraqi journalist hurl two shoes at President Bush?' or 'What public event occurred on 11 October 2002?' They were also able, when given a random computer-generated date, to say which day of the week this was and to recall a personal experience and a public event that occurred then. The public event was verified using a search engine.

It is amazing to think that someone who can remember that on 19 October 1987 there was a big stock market crash and the cellist Jacqueline du Pré died could have flaws in their memory. But even these HSAM people were susceptible to false memory implants. In one experiment the terrorist attacks of 11 September were talked about and footage of the United Flight 93 crashing in Pennsylvania was mentioned. Later, around 20 per cent of people recalled seeing this footage (both normal and HSAM people recalled this). One HSAM subject recalled: 'It just seemed like something was falling out of the sky.' In various other tests the HSAM and normal memory subjects experienced a similar level of memory distortions.

Bryan Strange from UCL demonstrated the power of emotion in relation to memory in a very simple experiment. His team showed that people were more likely to remember emotionally powerful words such as 'murder' or 'scream' than neutral words. The words that were forgotten most were those shown just before the emotionally charged ones. In this particular experiment the effect was greater in women.

How is it useful to managers who coach?

When you are leading team dynamics it is useful for everyone to be aware that memories cannot always be relied upon. Presume the best in your colleagues, rather than choose to believe they are lying. Consider ways that you can summarize meetings or projects that implants and strengthens positive, helpful memories. You have regular opportunities to help build supportive narratives with your people. Draw their attention to what went well. To the characteristics they showed that helped them and others and you want them to continue to invest in.

What can I do with a client or colleague this afternoon?

- Teach your clients about false memories – empower them to know that what they remember isn't always what occurred, so others may have different recollections – and these also may be distorted!
- Coach them through how they could explain this to others so they too may benefit from understanding how memories actually work.

The real value from understanding false memories is not likely to occur for you this very afternoon. The opportunities to be aware of this could spring up at any time with people you work with. Being able to offer a reason why people remember different things or what may underpin disputes in relationships is when this knowledge could help you to be even more effective.

Trust 38

What is it?

Trust can be defined as a strong belief in the reliability of someone or something. It can have both emotional and logical components to it. It is possible to get a gut feel that you can trust someone, what they say and will do. It is also possible to rationalize why you should or should not trust someone.

Why is it important for coaches/managers?

Stephen Covey makes some great points about trust and these are important for both personal and business coaches:

> When trust is low, in a company or in a relationship, it places a hidden 'tax' on every transaction: every communication, every interaction, every strategy, every decision is taxed, bringing speed down and sending costs up. My experience is that significant distrust doubles the cost of doing business and triples the time it takes to get things done.

(Covey, 2014)Most of us experientially know that trust is important. What neuroscience has done, again, is help us to understand how it is important from a very basic level. It adds another piece of the puzzle and enables us to organize ourselves and our workplaces with confidence.

In an economy and world where people are often encouraged to be independent and succeed, even to the detriment of others, this research and understanding is important. As coaches and managers we need to be equipped with as much quality information about how we work as possible so that we can operate beyond cultural changes, fads and ungrounded ideas. Incidentally, Paul Zak (2012) found that the countries with the highest trust levels are also the most economically successful.

Of course the other vital reason for people to be aware of the importance and effects of trust is so they can choose to invest appropriately in being trustworthy and building trust with those they work with. In corporate environments where external coaches may have people other than their client to report to this obviously needs to be handled thoughtfully. Most people who start to work with someone in a coaching role – be it a life coach or business-coach-type relationship, or more of a manager as coach scenarios are opening up – could experience vulnerability. The level of trust you can build from the outset can serve you well as the relationship progresses.

What is the underpinning neuroscience?

When we trust someone there is a knock-on effect on specific areas of the brain. The amygdala's activity levels are decreased. Threat and fear responses are low. On the other hand, if trust is breached then the anterior cingulate cortex detects conflict and activates the amygdala. We know that when there is threat and fear this obscures the brain's ability to do other things. Our ability to plan, to make quality decisions and to be creative all take a nosedive. Conversely, when trust is present our brain is freed up to engage in these things competently again.

Oxytocin is known to be important in being trustworthy, in choosing to trust others and feeling trusted. When we feel trusted we can actually become more trustworthy as a result of increased oxytocin levels.

It is worth noting that high levels of stress blocks oxytocin release. Evolutionarily this makes sense. If you are in a stressful situation your attention and focus needs to be on surviving. Oxytocin is important in being empathetic, relating to others' needs, and also has an effect on the area of the brain called the amygdala, which would diminish our ability to register anxiety. On a practical level this may mean waiting until someone calms down a little bit before offering them that hug. It also could be the reason that people can feel disconnected during times of stress, and why we may need to work extra hard to connect with them.

Another thing to be aware of when we're working to increase trust by increasing levels of oxytocin is that it can increase emotional pain. Oxytocin can strengthen social memories, meaning that a stressful situation can have a longer-term impact, triggering fear and anxiety in the future. This doesn't mean that building trustworthy relationships isn't a good idea (far from it). It is another component to be aware of, however. The mode of action here, the way that oxytocin intensifies stressful memories (the activation of the lateral septum), leads us to suspect that the same could be the case for strongly positive situations.

If we trust people unconditionally then the area of the brain called the septal area appears to be involved. However, if we trust someone based on certain conditions then the ventral tegmental area (involved in reward circuitry) is activated. This would be a scenario like someone arriving on time for an appointment, sending you a report you've asked for, or remembering to pick up dinner.

What are any interesting studies in this area?

The trust game is a classic research tool that has been played many times all over the world and studied by many different disciplines. Many studies involve the use of it, so an appreciation of the core game is useful.

The way the game works is that a subject sits at a computer screen and it is confirmed to them that they have received a sum of money just for showing up at the experiment. Let's say £10: this is now their money to keep. The subject (player A) is informed that the computer will ask another randomly chosen, anonymous player (player B) if they would like to transfer some or all of their £10 to player A. Why would player B do this? Well, because, as both players know, any amount that player B transfers to player A will be tripled and this will be added when the money hits their account. Player A will then be asked if they want to transfer any money to player B – remember that everything in the game is completely anonymous.

So this game has given rise to some fascinating studies. It is very interesting to see whether people act in both players' best interests, where player B gives their whole £10, resulting in player A having £40, and player A then giving £20 back to player B. Or whether, as economic theory used to predict, that player A should just keep whatever is given to them by player B and give nothing back.

Paul Zak led experiments that utilize the trust game. During one of these experiments player A is left with an average of $14 (in the United States) and player B with $17. They found that there was a 'dramatic and direct correlation between a person's level of oxytocin and her willingness to respond to a sign of trust by giving back real money' (Zak, 2012). There was even a correlation between the size of the transfer and the size of the recipient's response (more money meant more oxytocin). This oxytocin has a lot of impact, as we have explored in Chapter 12.

In another study, reciprocal altruism was studied using the 'prisoner's dilemma' game. For this, fMRI scanners were used to look at women's brains as they played the game, a version based on the classic scenario formalized by Albert W Tucker, a version of which is:

> Two members of a criminal gang are arrested and imprisoned. Each prisoner is in solitary confinement with no means of speaking to or exchanging messages with the other. The police admit they don't have enough evidence to convict the pair on the principal charge. They plan to sentence both to a year in prison on a lesser charge. Simultaneously, the police offer each prisoner a Faustian bargain. If he testifies against his partner, he will go free while the partner will get three years in prison on the main charge. Oh, yes, there is a catch... If *both* prisoners testify against each other, both will be sentenced to two years in jail.

The results showed that when there was mutual cooperation, areas of the brain involved in reward processing were activated: nucleus accumbens, caudate nucleus, ventromedial frontal cortex, rostral anterior cingulate cortex. The experimenters proposed that the activation of this neural network positively reinforces reciprocal altruism... trust is being built! A later study showed that when cooperation wasn't reciprocated – so a version of when your partner testifies against you – the amygdala and insular show increased activity and people report feeling angry, disappointed and irritated.

One study suggested that decisions about trustworthiness may involve both deliberate and emotional evaluations. The experiment involved participants going into an fMRI scanner and being shown pictures of faces. The first task they were then asked to perform was to indicate whether the face looked trustworthy or not. The second task involved the participants indicating whether the face looked like that of a high school student or a college student. This is known as performing an unrelated cognitive evaluation. Their brains were scanned while they were doing these tasks. There was a little bit more to the study design to make it really meaningful. The results showed that the amygdala was activated more when people saw the untrustworthy faces compared to when they saw the trustworthy ones. The same was noted with the right insula and fusiform gyrus. In another study that looked at people with lesions in the amygdala it was found that those people had impaired judgement of untrustworthiness. So these two studies tie together and help build a reliable picture. The other task that the subjects performed gave a contrast frame. This enabled the researchers to see that the right superior temporal sulcus was activated more when judgements about trustworthiness were being made rather than judgements about age. The orbitofrontal cortex was also involved.

The suggested conclusion from these experiments is that visually relevant information is processed by the superior temporal sulcus. Then the amygdala and orbitofrontal cortex create the emotional response. The amygdala appears to be responsible for the rapid and automatic emotional responses. On the other hand, the orbitofrontal cortex appears to contribute only in the context of a particular conscious evaluation. The emotional response people experience could include both cognitive and somatic changes. We tend to say that 'we have a feeling about someone' as a result.

Our final study in this chapter looks at the effect of feedback on behaviour. The experiment involved participants making decisions about whether or not to trust trading partners after reading descriptions about their life events that were either slanted towards being praiseworthy, neutral or suspect moral characters. Normally the striatum is involved in us learning and adjusting future behaviours based on reward feedback. In this experiment the participants had to make risky choices and, very interestingly, they persisted in making more risky choices with the partner who had a praiseworthy character. The way the study was designed meant that each of the partners (praiseworthy, neutral and suspect) equally were reinforced. We know from other studies that the caudate nucleus's activation differentiates between positive and negative feedback. In this case it only held true for the neutral partner; it did not at all for the praiseworthy character and only did weakly for the suspect one. The suggested conclusion here is that previous social and moral perceptions can diminish our reliance on feedback mechanisms in the neural circuitry of trial-and-error reward learning.

How is it useful to managers who coach?

Trust is essential to create High-Performing Neural Environments. If people aren't safe, then they can't fully dedicate their cognitive resources to those higher levels of processing. A big part of a manager's ongoing KPI's should be to create a high trust environment. They need to be aware that trust can take a long time to build but be quick to damage. There is a lot of advice around how to build trust, many good ideas out there. We'd suggest you think about someone you care deeply about and want the best for. Use this as your barometer for whether your action is treating your colleague well, or whether it could be improved. Would you want this person's manager treating them in that way?

What can I do with a client or colleague this afternoon?

- Raising the awareness of the importance of trust and the implications of trust can help your clients deal with various situations.

- Ask your client how they feel about trust and being trustworthy. Explore how they think people naturally behave.

- Task your client to make a list of the top 10 people who they want to feel they are trusted by. Then suggest they make a list of all the ways they can build trust with those individuals. For example, if one of those people is a direct report of theirs at work perhaps your client will want to choose one thing a week for four weeks that they will tell their colleague that they will do and by when. Then (and here is the revolutionary part for some people) they ensure that they do it.

In a training event with MDs once, an intelligent man questioned whether we could build trust. Here we take the responsible approach that you can and should. While the understanding of what trust is may develop further, currently it is fair to suggest that the neural networks that are laid down over time are important. We are in control of taking actions that others view as trustworthy – and so we are in control of building our reputation as a trustworthy individual.

Fairness 39

What is it?

Definitions of fairness include an element of being legitimately sought, or achieved, being proper under the rules, and free from bias and dishonesty, along with many more. Most people have a sense of fairness that is socially and contextually programmed in.

Why is it important for coaches/managers?

We do not like it if we don't perceive things to be fair. Our response can be a big one and can have lots of knock-on effects. Being aware of this inbuilt response as a coach or manager is very important. Taking your questions in a certain direction is valuable if you suspect the perception of fairness to be a possible cause of challenge. There may also be times when you will want to explain the neuroscience of the fairness or unfairness response to the person you are working with so that they can help others or manage situations more effectively.

What is the underpinning neuroscience?

The anterior cingulate cortex is involved in the process of evaluating whether outcomes are met or are different from expectations. It is also involved when we suffer social losses, such as being rejected or being treated unfairly.

The insula has been shown to light up when people are averse to inequity. People have rejected monetary rewards if they think they are unfair relative to the rewards of others. Research is leading us to believe that, in order for people to accept unfairness, the prefrontal cortex (PFC) is very important. At the moment it looks as if the cognitive PFC quiets down the emotional gut-feeling part of the brain and overrides it.

Productivity takes a hit when people perceive unfairness, because there is decreased activation of the reward centres in the brain and the emotional parts of the brain are disrupted along with the cognitive areas!

What are any interesting studies in this area?

In the area of fairness within social behaviour there is an experimenter's game that is often used. It is called the ultimatum game. The way that humans play the ultimatum game flies in the face of old economic theory. We would expect people to behave in a way that is self-serving. If I offer you £2 (no strings attached) then I would expect you to take it. However, if you found out that someone had given me £10 and I was only choosing to share £2 with you then it becomes a different matter. The ultimatum game has been played all over the world in different currencies, with different amounts, with people of different cultures and the basic results stay the same. Unfair offers are rejected.

A study by Tabibnia and Lieberman (2007) illustrated some important points. Today many organizations offer monetary incentives to try to motivate their employees and then wonder why things don't go exactly to plan. We know from the field of social cognitive neuroscience that human behaviour is far more complex. This study showed the importance of fairness. In the study they saw that fair offers led to both higher happiness ratings and also increased activity in several reward regions of the brain when compared with unfair offers of the same monetary value. Specifically, 50 cents generated a bigger neurological reward than $10 – which sounds crazy until you know that it was 50 cents out of $1, and $10 out of $50.

In another study, this time by Singer *et al* (2006), male and female volunteers played an economic game. Partners played either fairly or unfairly. The volunteers had their brains scanned with fMRI while watching the partners receiving pain. Both men and women exhibited empathy-related activation in the fronto-insular and anterior cingulate cortices towards their partners who had played fairly. In men this response was quite different towards those partners who acted unfairly. The empathy response was reduced and there was an increased activation in the reward-related areas of the brain.

Another study with Tania Singer *et al* got men and women to play an economic game, set up to play either fairly or unfairly. During the second part of the experiment the participants' brain activity was measured using fMRI. For this part of the study the participants were able to see their pair receiving pain. In both men and women the fronto-insular and anterior cingulate cortex was activated, signalling empathy, perhaps, towards the fair players. When pain was inflicted upon the unfair players, however, the empathy-related responses were significantly reduced in men. Instead, there was activation in the reward-related areas of the brain, perhaps signalling a desire for revenge. So the consequences of what is considered fair are significant, especially for men.

In another study participants had their electroencephalogram (EEG) recorded while they played an ultimatum game. They were the recipients during the game and so responded to both fair and unfair offers from other human (so not a computer) proposers. The participants also rated their concern for fairness while they went through this process. Currently it is believed that something called medial frontal negativity (MFN) is linked to the anterior cingulate cortex (ACC), which we know is involved in unfairness. This experiment showed that MFN was more pronounced after unfair offers were made than when fair offers were made. This was even more the case for the individuals who self-reported high concerns for fairness. This research suggests that MFN reflects both whether outcomes match expectations and also whether the process by which these outcomes occur match social norms. The research also suggested that people in bargaining situations are concerned with the fairness of an interaction, not just the outcome for themselves.

Obviously this is only one study, and one that puts people in an artificial situation. So this research should not be extrapolated to mean that every person who has to negotiate as part of their work would have challenges. This is not what the research is saying.

In a further study that expands our understanding of the MFN's response within the social brain, the participants in the study played a classic competitive gambling game where in order for one person to gain, the other had to lose. EEG was employed to measure the neural activity of the individual's perceptions. The MFN showed a significant gender difference. When a female player's opponent lost out in the game the MFN was elicited in them. This can be interpreted that they categorized this occurrence as negative, even though it was to their financial benefit. The same thing did not happen in men.

At a fundamental level, how fairly an individual perceives a situation – be that a work scenario, a family incident or an issue with friends – can impact the outcome of that situation. One study looked at the neural segregation of objective and contextual aspects of fairness. They used a modified version of the ultimatum game and varied the social context. They induced a bias in the participants' acceptance of objectively identical offers and used fMRI to explore correlations between objective and contextual aspects of fairness and the brain's response to these. Their findings show that objective social inequality is tracked in the posterior insula cortex. This objective inequality is integrated with social context in posterior and mid-insula. This is consistent with a fairness construct that is adapted to the social environment. The researchers propose that this might help to explain why we see seemingly inconsistent responses to things that happen. Sometimes people seem to let unfairness slide, and it is not an issue. At other times these same people find a similar (objective) situation to be completely unacceptable. Being aware that fairness is processed by integrating objective and contextual aspects sheds light on why this might be.

An experiment from a different angle looked at how we limit any selfish motives we may have and how we implement fair behaviour. The experiment doesn't address the whole big question (there is a lot to this area in brain terms) but does offer some insights. The study documents the results of disrupting the right dorsolateral prefrontal cortex (dlPFC) (using low-frequency repetitive transcranial magnetic stimulation). The researchers found that this reduced the participants' willingness to reject unfair offers. This indicates that the right dlPFC plays an important role in implementing fair behaviours. Could this mean that if our PFC is not working optimally for whatever reason that tendency to behave fairly may be compromised? Perhaps.

Our final study in this chapter is an uplifting one. It explores the positive emotional impact of fairness. The study examined the neural responses to fair and unfair offers and self-reported happiness levels while controlling for monetary payoff. This is important because, unsurprisingly, the mental processes underlying preference for fairness – and for preferring more money – need to be separated. The clever researchers looked at what happened when participants were given unfair offers that were of equal monetary value to fair offers. They found that high happiness ratings were associated with the fair offers and activation of several reward regions in the brain. They also found that the tendency to accept unfair proposals was accompanied by increased activity in the right ventrolateral prefrontal cortex. This part of the brain is involved with emotion regulation. There was also decreased activity in the anterior insula (involved in negative effect). So, taken all together, we can believe that fairness is hedonically valued. Also, that when we have to tolerate unfair treatment for material gain our brain is going through processes similar to suppressing negative effect.

How is it useful to managers who coach?

Organizations need to be very mindful of our need for perceived fairness. Often the response by individuals will not be a public one if they detect unfairness. As a leader or a manager you almost need to develop a second sense for what could possibly be interpreted as unfair, and pre-empt it either with great communication (if it isn't really unfair) or by making adjustments. Another approach could be to invest in a culture of transparency and trust so that people feel able to voice when they think things are unfair so that they can be openly addressed.

As a manager you could consider sitting down with your team and drawing attention to all the fair things that already exist within the organization. Sharing how decisions are made for things that are important to them – for example promotions, pay rises, who gets to sit where in the office, etc – can help reduce any unconscious tendencies.

A slightly left field, but very valid, application of the concept of fairness is to consider how an organization is positioning itself to the outside world. An organization whose activities are perceived as unfair, for example the actions Nestlé took many years back in distributing baby milk for free in countries where people could not then afford to buy it, took a long time for people to forgive.

What can I do with a client or colleague this afternoon?

The application of this depends heavily on context. Here are some examples, though:

- If a client has identified a lack of perceived fairness at work then being equipped to have a conversation about the impact that has on people (even citing research) could be a great first step.
- If a client is managing others (in any setting, from business to children to a social group) they will need to be aware of creating transparency and having clear frequent communication.
- Creating clear expectations is another very important thing that you can both use with your client (contracts, agreements, check-ins), and also to encourage your client to evaluate how effectively they are using fairness in their various interactions with people.

Loneliness 40

What is it?

Some definitions of loneliness say that it is the state of being alone in solitary isolation, a universal human emotion or a complex set of feelings encompassing reactions to unfulfilled social needs. In this context, loneliness can definitely be experienced even when physically with others. In the terms through which we will explore it here, we can think of it as social connectedness. It is subjective, so a person may have a family, friends and colleagues yet still feel lonely: they can still have a sense of an absence of connectedness. At times it is normal to feel a fleeting sense of loneliness, to feel 'out of it' for example at a party or social gathering. It is not uncommon to feel on the periphery of things from time to time. The challenge, and our focus here, is on when those feelings are prolonged: chronic loneliness. From what we know of the brain, this is likely to have a big impact on individuals.

Why is it important for coaches/managers?

When people are lonely they can struggle in many areas of their life. Their performance at work is likely to suffer (cognitive abilities are affected). They are likely to be less happy at home. Their friendships and relationships could deteriorate. Their health is likely to suffer and self-restraint is affected. Since people can feel lonely even when they seem to have lots of people around them whom they have good relationships with, it is important to be mindful of this.

It has been shown that individuals suffering from chronic loneliness had elevated salivary cortisol levels throughout the course of a day. This suggests that there is increased activation of the HPA axis.

Cacioppo *et al* (2009) suggests that there are three complex factors that are very important when we explore the effects of loneliness. The first is how vulnerable we are to social disconnection. Genetically we are predisposed to need a certain amount of social inclusiveness. Linked to this, as is frequently the case, the environment we grew up in and continue to exist in has a big role to play with how our DNA expresses itself. So there is a degree of where our client is at – that has been predetermined by the time we work with them. Second, there are all the factors that affect a person's ability to self-regulate how they respond to feelings of isolation. An important component here is not just the outward response that people have but the internal one as well. Third, our mental perception of what is happening. How we interpret what is occurring is obviously also fundamental.

What is the underpinning neuroscience?

Dr Ryota Kanai from UCL Institute of Cognitive Neuroscience says that there is a neurobiological basis for loneliness and that through training people may be able to improve their social perception and become less lonely. After doing 108 brain scans Kanai *et al* (2012) found that the lonely people had less grey matter in the left posterior superior temporal sulcus. This area is implicated in basic social perception. In a very simple test of perception – judging misaligned eyes on faces and whether the people were looking left or right – the lonely people found the test to be much harder.

This is not the entire story when it comes to the topic of loneliness, but another interesting insight into the correlation between what is happening in our lives and the plastic nature of our brain.

What are any interesting studies in this area?

It is worth bearing in mind a study that showed it was possible to *create* feelings of loneliness in people. This study took a group of individuals and manipulated their perceived levels of loneliness with the help of hypnotherapy. The psychiatrist doing the hypnosis used a precise script to reconnect the participants to memories that would make them feel either very connected or very lonely. They found that they could fundamentally alter the responses given to the UCLA Loneliness Scale test. It would be unethical to leave people in a place where they felt lonely. It does open up questions for potential help, though. What would happen to the brains of the individuals made to feel lonely? What would happen to the brains of the individuals made to feel connected?

A study by Lieberman *et al* (2003) – 'Does rejection hurt? An fMRI study of social exclusion' – really highlighted the effect of social pain on the brain. In this classic experiment individuals were scanned while they played a virtual ball-throwing game with others (or so they thought). After a while they were excluded while the people they played with kept throwing it between themselves. They found that the dorsal anterior cingulate cortex was activated more when they were excluded. The participants also reported feeling distressed. This is a similar response we see in the brain to when someone is in physical pain. The right ventral prefrontal cortex (vPFC) was also active during exclusion. The more it was activated the less distress people reported. This, along with the ACC changes mediating the right vPFC-distress correlation, leads us to think that the cognitive PFC is involved in regulating the distress by disrupting ACC activity.

In the 1990s John Cacioppo did an experiment using the UCLA Loneliness Scale to separate students into three groups (Cacioppo *et al*, 2000): the 'very lonely', the 'not at all lonely' and the ones in the middle. These students then went through a dichotic listening cognitive test. The experiment was a clever one utilizing the brain's natural dominance in one, normally the left, hemisphere. Each student wore a headset and each was trying to identify a sound (like 'wu' or 'ha') that came into one ear, while trying to ignore the irrelevant sound coming into the other ear. Without any additional instructions people showed right ear preference (as expected). When told to focus on using their right ear each group did so with similar success. It got interesting when they were asked to focus on their non-dominant ear. This takes more effort. The group with normal levels of loneliness and those who weren't at all lonely still did well on this test. However, the lonely ones didn't perform as well. They struggled to impose this conscious control and were less accurate in their results.

A couple of related experiments that produced fascinating results were led by Roy Baumeister (Baumeister *et al*, 2005). They found that when people were manipulated to believe that they were likely to end up alone (versus a happy future with significant others) they persevered in drinking more of a concoction. This concoction was positioned to them as being unpleasant in taste but healthy and nutritious. The related study looked at how well people would be able to regulate a behaviour that felt good but wasn't good for them. This time individual participants were told that they would need to make small groups with people who liked and respected each other. After some mingling time they were asked who they wanted to work with. Some individuals were then told that unfortunately no one had selected to work with them, but that they could just do the next part of the task alone. The other individuals were told that because everyone had wanted to work with them, and it was impossible to have such a big group, they would have to do the next part of the task alone. They were told that they were doing a taste test and were presented with 35 bite-sized chocolate-chip cookies in a bowl. They were asked to test them by eating as many as they needed to make an accurate judgement of taste, texture and smell. The disconnected participants (the ones no one wanted to work with) ate an average of nine. The connected participants ate roughly half of that. The first group also reported that the cookies tasted better... but still only mediocre!

The neural correlates for this relationship between social exclusion and self-control were explored with the help of a magnetoencephalography (MEG) investigation. MEG was used with the participants while they did a task where they had to solve 180 medium-difficulty mathematical problems – this was designed to assess their self-control. Some participants were primed to feel socially excluded while the others were in the control group. The socially excluded people showed less activity in the occipital and parietal cortex and the right prefrontal cortex. The results suggest that social exclusion interferes with the executive control of attention. This can be noticed when people are performing specific components of cognitive tasks.

Considering what can be done to help people who feel lonely is a complex task. An experiment that gives us another important insight into what is actually happening was carried out at the University of Chicago. Volunteers underwent fMRI and were shown photographs during this process. Some of the photos were of people, some of objects. Many of the photos were chosen to evoke a positive or negative emotional impact. After the scan, the participants' levels of loneliness were measured. The people who were not lonely showed a greater response to positive pictures of people than positive pictures of objects. There was greater activity in the limbic region, specifically the ventral striatum (known as a reward centre). It is very interesting that the pictures of people had different effects on the lonely versus the non-lonely people. We can draw from this that the answer is not as simple as just putting lonely people in situations with other people (if we hadn't already guessed that!).

Equally interestingly, when the lonely people were shown a negative photograph of people they paid more attention to this than the negative photograph of an object. Their visual cortex and temporoparietal junction showed increased activation.

In the classic experiment where one person acts as 'proposer' and the other as 'decider' and the proposer is given £10, lonely people performed differently to the non-lonely. The proposer gets to choose how to split the money with the decider. The decider either rejects the proposal (in which case neither person gets any money) or it is accepted in which case both get to keep the money. Fairness is key to this game. In this version of the game the propositions were rigged so that half of the offers were unfair (less than £3 proposed). Interestingly, the lonely people accepted more unfair offers than the non-lonely people. They were comparable in their rating of the offers (so they recognized that some of the offers they were accepting were unfair) but accepted them anyway.

Another study looked at the mechanisms by which loneliness might have a negative effect on health. It focused on health behaviours, cardiovascular activation, cortisol levels and sleep. In this particular study, of which there were some limitations (there almost always are), they found that both cardiovascular activation and sleep function were compromised in lonely people.

How is it useful to managers who coach?

Loneliness is a silent killer. It negatively affects both physical and mental health and the fact that it is often hidden makes it even more dangerous. Gone are the days of the only lonely person being the old person who lives on their own. People who are affected by this can be married, working physically side by side with colleagues, have friends and have children. Yet, still feel alone. As a manager there are several things that it is your responsibility to do:

- Get alongside people and develop real relationships with them.
- Facilitate positive interactions between team members.
- Create opportunities for individuals to contribute (and be recognized for that contribution).
- Help people connect to the higher purpose of their work.
- Encourage team gratitude, compassion and empathy.

What can I do with a client or colleague this afternoon?

Social connectedness is key to people's performance in life. Therefore, assessing and investing in increasing this can be very valuable. Consider:

- exploring what loneliness really is with your client;
- identifying ways they think they could cushion against it;
- taking the UCLA Loneliness Scale test;
- looking at undesirable behaviours and exploring whether increasing social connectedness would help;
- supporting your client to improve their social perceptiveness;
- drawing up a plan to increase connectedness;
- helping your client to evaluate their current relationships – the quality rather than quantity of them.

Check out my LinkedIn Learning course specifically on Loneliness in the Workplace.

Conclusion

This book is the first stepping stone of several in any coach's journey to really understanding the neuroscience that is useful to us. It has introduced you to the brain areas that are of most interest and why that is the case. Then we concentrated on the chemicals that will help you to understand the thoughts and actions of your client or colleague. Setting aside the neuroanatomy and neurobiology we looked at the foundational concepts that would crop up frequently as we moved forward, including the pivotal neuroplasticity. Drawing on our neuroanatomy we took a different approach to widen our understanding and looked at brain networks. Like watching a good play, we paused for an intermission – where we played with the quantum brain. The final chunks of our exploration were focused around the classic and not-so-classic areas of coaching, where there have been fascinating studies performed that can strengthen and focus our work. The previous editions of this book have already been quoted in other books. It's an exciting time.

If you have been interested, inspired or excited by anything that neuroscience is offering the world of coaching I would encourage you to take the next step. We have the opportunity to be more credible and more evidence-based and subsequently serve more people better.

By offering insights from neuroscience as an underpinning to coaching we don't have a static model that we are limited by. This is exciting because the field is growing and developing at such a rate it is likely that a revision of this book in five years' time would be double its size! When coaches walk into our Neuroscience for Coaches programme or managers join our Coach Savvy Manager programme my aim for them is to develop their thinking. There are ways to do that – and encouraging playing with ideas, a safe environment in which to explore ideas and a variety of different sources have proved good starting points. Graduates of the programme are invited to continue to stay up to date by joining the NS4C Club. This dedicated group of coaches will grow and explore the applications of neuroscience to our field. My hope is that together we can be pioneers in shaping this for future generations.

The future of coaching

Speaking to amazing coaches around the world highlights how the coaching profession has changed over the years. It continues to evolve.

I asked Susan Grandfield: 'What do you think the future holds for coaching, neuroscience and organizations? Do you think things will change?'

> I think things will change, or maybe not change so much as continue to evolve. Coaching absolutely has evolved from the way it first started. I've been coaching for 10 years now and people get what coaching is more now than they did 10 years ago. I think people can see the benefits of coaching so the thing that might change most is within organizations. Development programmes that incorporate some level of one-to-one coaching for each individual create the big shifts for people. Learning new stuff in a group environment is really helpful, but then in terms of 'How do I actually apply that to me and my situation and get the benefit from it', the one-to-one coaching really helps. I think the Chartered Institute of Personnel and Development (CIPD) talk about the increase in coaching, but coaches are becoming more and more a feature of organization learning and development and organizational development itself. So I think coaching will continue to evolve.

When I asked Christian what he thought might be on the horizon for the future in coaching he had a reply congruent with his area of academic focus.

> I ran a workshop recently called 'Towards the philosophy of coaching' at the big Positive Psychology conference. It was oversubscribed and we had to run the workshop again, so there was a lot of interest. One of the first questions I asked was 'How many of you are coming with a kind of a coaching perspective?' Half the room raised their hands. Then I said: 'How many of you are coming from a positive psychology angle to this?' and then the other half raised their hands. I guess my own desired vision of the future would be that that becomes a silly question, that people go: 'What kind of question is that? Of course we're from both. We study both.' The area of interest of positive psychology, which is how we get to optimal human functioning and how we bring science into understanding well-being, is the area coaching wants to be working in. I think the area of interest for both positive psychology and coaching is what Jim Collins calls 'the good-to-great bit'. Coaching and positive psychology both have a lot to offer each other. For example coaching can offer a lot to positive psychology practitioners because it's very practice-based, it's all about interpersonal relationships and practical ways of supporting people to flourish one to one. Then, equally, positive psychology has a lot to teach coaching

because it can be applied across organizations so they're thinking more strategically, and you can use the interventions with groups of people across countries. So I guess the future for me is integration so that we've got multi-skilled, multidisciplinary professionals who are able to support individuals and organizations to flourish.

For us, the future will involve people – from toddlers to the very old – knowing more about how their brains work and how they can live more fully.

COACHES IN THE KNOW

We know that keeping up to date with the latest thinking is important. But with so much information out there, it's hard to know where to start. Our Neuroscience for Coaches community is a place to discover cutting-edge insights drawn from the many neuroscientists, psychologists and interesting people who are working in this field and who we think are worth paying attention to. The research they do, the papers they publish and the books they write are of high quality and deliver fresh insights. By checking in with our blogs and videos you'll keep your knowledge up to date and stay credible.

Visit www.neuroscienceforcoaches.com to join our community and we'll let you know when there are new resources for you to take advantage of.

Want to go deeper?

We run programmes both publicly and in house so check out what would be the best fit for you at:
www.neuroscienceforcoaches.com/programmes

We'd also love you to join our LinkedIn group: Neuroscience for Coaches Global

CALLING ALL HR PROFESSIONALS

When it comes to people development, your managers on the ground are key to getting the most out of your people. It therefore makes sense that managers with people development responsibilities have much to gain from a coaching-style approach.

At Synaptic Potential, we run workshops and training programmes that bridge the divide between management and coaching – equipping your managers with the latest skills, mindsets and behaviors they need to coach. You can find out more here: https://synapticpotential.com/solutions/managers/the-coach-savvy-manager

GLOSSARY

action potential   The electrical spike that occurs within neurons by way of communicating with other cells, often referred to as 'firing'.

adrenaline   Well known for its role in the fight-or-flight response and as being part of our emotional response to things.

amygdala   The almond-shaped amygdala is deep within the brain connected to the hippocampus. It is heavily involved in emotional regulation.

anterior cingulate cortex (ACC)   The area of the brain primarily responsible for conflict or error detection.

axon   A nerve fibre that conducts electrical impulses.

basal ganglia   There are billions of basal ganglia, located deep inside the brain and these are key to storing routines, repetitive behaviours and thoughts – so are associated with routine behaviours; habits.

cerebral cortex   The outer layer of the brain.

cognition   Mental processing, including attention of working memory, problem solving, decision making and reasoning.

cognitive social neuroscience   The interdisciplinary field that seeks to understand how biological systems, like humans, implement social processes and behaviours. It uses imaging tools that cognitive neuroscientists favour, such as MRI scanners.

cortisol   A hormone that gets released during times of stress. It does several things, including increasing levels of blood sugar and suppressing the immune system.

dendrites   The branches at the cell-body end of a neuron that propagate the electrochemical stimulation received from other cells.

dopamine   The neurotransmitter involved in how interested we are in things, reward and pleasure networks, regulating emotional responses. This is important in learning.

electroencephalography (EEG)   The process where people wear what looks like a swimming hat with wires that records electrical activity from the scalp.

emotions   Different definitions in different fields, perhaps useful to consider here is an experience that correlates with a particular physiological activity pattern.

episodic memory   The memory of autobiographical events, they have an emotional component to them and may trigger behavioural changes.

excitation   The chemical process that makes a neuron more likely to fire an action potential.

feelings   The psychological definition is normally the conscious subjective experience of emotion.

fight-or-flight response   Typically when a presumed threat is present the sympathetic nervous system floods the body with adrenaline and noradrenaline, which raise the heart rate, breathing rate and prepare the body to 'fight' or run away.

flow   The term coined by Mihaly Csikszentmihalyi for a specific state of effortless concentration, enjoyment, energized involvement and success in the progressing of the task with which you are involved.

frontal lobes   The area at the front of the brain, rich in dopamine-sensitive neurons. Very important in higher mental functions, including executive functions.

functional magnetic resonance imaging (fMRI)   A type of scanning process used predominantly in research fields that measures brain activity by detecting changes in blood flow. It is similar to MRI (used mostly diagnostically in the clinical world) except that it uses the change in magnetization between oxygen-rich and oxygen-poor blood.

glial cells   Cells that maintain balance, form myelin and provide support and protection for neurons.

Hebb's law   The theory that is summarized as 'cells that fire together, wire together' and is the mechanism for neuroplasticity.

hemispheres   The brain can be seen to have two sections divided by a groove; each section is called a hemisphere. They are connected by the corpus callosum, a large bundle of nerve fibres.

hippocampus   Brain area involved in memory storage, including emotional memories.

hypothalamic-pituitary-adrenal (HPA) axis   The influences and feedback processes between the hypothalamus, pituitary gland and adrenal glands. Forms part of the neuroendocrine system that controls responses to stress.

hypothalamus   An almond-sized structure buried deep inside the brain. It is responsible for regulating several things and controlling their linked behaviour, such as eating, sexual behaviour, sleeping, body temperature, our emotions and the secretion of hormones.

inhibit   In cognitive terms it is the mind's ability to tune out irrelevant distractions.

insular cortex (or insula)   Enables us a degree of awareness and is involved in our ability to be 'in tune' with ourselves.

limbic system   A controversial term (some neuroscientists think it should be abandoned) used to refer to a group of brain structures, including the hippocampus and amygdala, that are involved in a variety of things including emotion, behaviour and long-term memory.

mind   Possibly arising from brain functions; considered the cognitive faculties that enable consciousness and thinking.

myelin   The electrically insulating material that forms a sheath around the axon of a neuron, essential for proper functioning.

neocortex   The outer layer of the cerebral cortex, six layers thin, involved in higher functions.

neuroimaging   The group of technological techniques that enable us to directly or indirectly image the structure, function and pharmacology of the brain.

neuron   Core cells of the nervous system, electrically excitable and transmit information through electrical and chemical signals.

neuroplasticity   The term to describe the adaptable nature of the brain, both on a micro and macro level.

neuroscience   The scientific study of the nervous system. Now an interdisciplinary science that collaborates with other fields.

neurotransmitter   The chemicals that transmit signals from one neuron to another cell across a synapse.

noradrenaline   A compound that has many roles, including as a hormone and a neurotransmitter in the body.

nucleus accumbens (NA)   An area of the brain, part of the ventral striatum that is part of the basal ganglia. It is involved in pleasure, reward, motivation, rein-forcement learning, fear, addiction, impulsivity and the placebo effect.

orbitofrontal cortex   The area of the brain located within the prefrontal cortex in the frontal lobes. It is involved in the cognitive processing of decision making and linked to emotion, memory and reward.

oxytocin   Involved in social behaviour, increasing trust, decreasing fear, increasing generosity and also in cognitive functions.

prefrontal cortex (PFC)   Anatomically the frontmost part of the frontal lobes. Involved in executive functions, planning, decision making and social behaviour.

reuptake   The process by which a neurotransmitter is reabsorbed back into a presynaptic neuron. It facilitates the recycling of neurotransmitters and controls how long a signal from a neurotransmitter lasts.

serotonin   Important for mood regulation, appetite, sleep, memory and learning.

striatum   Both coordinates motivation with movement and is also involved with rewards and how we are motivated. It is part of the basal ganglia.

synapse   The structure that enables a neuron to communicate to another cell. The synaptic gap is the area between the neuron and the other cell.

thalamus   You can think of the thalamus as a relay station where nearly all the information filters through to go into the cortex (which is the outermost area of the brain). It is involved in pain sensation, attention and alertness.

triune brain   The now controversial model of the brain consisting of the reptilian, limbic and neocortex.

ventrolateral prefrontal cortex (vlPFC)   The part of the prefrontal cortex that is located opposite the back and away from the midline.

white matter   Composed mostly of glial cells and myelinated axons it modulates the communication between brain regions.

working memory   The limited-capacity memory system that holds information, normally for only a short time. Using it is energy intensive and requires the prefrontal cortex.

References

Introduction

Dilts, R (2003) *From Coach to Awakener*, Meta Publications, Capitola, CA
Klein, N (2002) *Time to Think: Listening to ignite the human mind*, Cassell, London

Part One

MacLean, P D (1990) *The Triune Brain in Evolution: Role in paleocerebral functions*, Springer, New York

Chapter 1

Bicks, L K *et al* (2015) Prefrontal Cortex and Social Cognition in Mouse and Man, *Frontiers in Psychology* 6, 1805

Carlson, R W and Crockett, M J (2018) The lateral prefrontal cortex and moral goal pursuit, *Current Opinion in Psychology* 24, 77–82

Dixon, M L (2017) Emotion and the Prefrontal Cortex: An Integrative Review, *Psychological Bulletin*, 143 (10) 1033–1081

Girotti, M *et al* (2018) Prefrontal cortex executive processes affected by stress in health and disease, Prog Neuropsychopharmacol Biol Psychiatry 85, 161–179

Hiser K M J (2018) The Multifaceted Role of the Ventromedial Prefrontal Cortex in Emotion, Decision Making, Social Cognition, and Psychopathology, *Biol Psychiatry* 83(8) 638–647

Journal of Experimental Psychology: Human Perception and Performance, 27 (4) 763–97

McEwen B S and Morrison, J H (2013) The brain on stress: vulnerability and plasticity of the prefrontal cortex over the life course, *Neuron* 79 (1) 16–29

Meyer, D E *et al* (2001) Executive control of cognitive processes in task switching, *Journal of Experimental Psychology: Human Perception and Performance*, 27 (4), pp 763–97

Nyberg, L (2018) Cognitive control in the prefrontal cortex: A central or distributed executive?, Special issue on Brain, mind and behavior, *Scandinavian Journal of Psychology*, 62–65

Paneri, S and Gregoriou, G G (2017) Top-Down Control of Visual Attention by the Prefrontal Cortex Functional Specialization and Long-Range Interactions, *Frontiers in Neuroscience* 11, 545

Chapter 2

Caligiore, D *et al* (2019) The super–learning hypothesis: Integrating learning processes across cortex, cerebellum and basal ganglia, *Neuroscience and Biobehavioural Reviews* 100, 19–34

Da Cunha, C *et al* (2012) The role of the basal ganglia in motivated behavior, *Rev Neurosci*, 23 (5–6) 747–767

Florio, T M *et al* (2018) The Basal Ganglia: More than just a switching device, *CNS Neurosci Ther* (8) 677–684

Harber, S N (2014) The place of dopamine in the cortico-basal ganglia circuit, *Neuroscience*, 12, 248–257

Hikosaka, O *et al* (2017) Paralel basal ganglia circuits for decision making, *J Neural Transm*, 125 (3) 515–529

Kim, H F and Hikosaka, O (2015) Parallel basal ganglia circuits for voluntary and automatic behaviour to reach rewards, *Brain*, 138, 1776–1800

Chapter 3

Axmacher, N *et al* (2010) Intracranial EEG correlates of expectancy and memory formation in the human hippocampus and nucleus accumbens, *Neuron*, 65 (4), pp 541–49

Baez–Mendoza, R and Schultz, W (2013) The role of the striatum in social behavior, *Frontiers in Neuroscience*, 7, 233

Bhanji, J P and Delgado, M R (2014) The Social Brain and Reward: Social Information Processing in the Human Striatum, *Wiley Interdiscip Rev Cogn Sci*, 5 (1) 61–73

Burton, A C *et al* (2015) From ventral–medial to dorsal–lateral striatum: Neural correlates of reward–guided decision–making, *Neurobiol Learn Mem*, 0, 51–59

Castro, D C and Br ropeptidergic Hub: Anatomical and Functional Diversity within Nucleus Accumbens Shell, *Neuron*, 102 (3) 529–552

Fareri, D S and Delgado, M R (2014) The importance of social rewards and social networks in the human brain, *Neuroscientist*, August 20 (4) 387–402

Fareri, D S and Delgado, M R (2014) The importance of social rewards and social networks in the human brain, *The Neuroscientist*, 21 February [Online] http://nro.sagepub.com/content/early/2014/02/21/1073858414521869 (archived at https://perma.cc/V2ZQ-Q4RV)

Floresco, S B (2014) The Nucleus accumbens: An Interface Between Cognition, Emotion, and Action, *Annu Rev Psychol*, 66, 201–2028

formation in the human hippocampus and nucleus accumbens, *Neuron*, 65 (4) 541–49

Graybiel, A M and Grafton, S T (2015) The Striatum: Where Skills and Habits Meet, *Cold Spring Harb Perspect Biol*, 7 (8) a021691

Harris, H N and Peng, Y B (2019) Evidence and explanation for the involvement of the nucleus accumbens in pain processing, *Neural Regen Res*, 15 (4) 597–605

Kim, B S and Im, H I (2018) The role of the dorsal striatum in choice impulsivity, *Annals of the New York Academy of Science*, 1451 (1) 92–111

Myslivecek, J (2021) Two Players in the Field: Hierarchical Model of Interaction between the Dopamine and Acetylcholine Signaling Systems in the Striatum, *Biomedicines*, 9, 25

Olds, J (1956) Pleasure centers in the brain, *Scientific American*, pp 105–16

Wenzel, J M *et al* (2015) A Role for Phasic Dopamine Release within the Nucleus Accumbens in Encoding Aversion: A Review of the Neurochemical Literature, *ACS Chem Neurosci*, 6 (1) 16–26

Chapter 4

Bellucci, G *et al* (2018) The role of the anterior insula in social norm compliance and enforcement: evidence from coordinate–based and functional connectivity meta-analysis, *Neurosci Biobehav Rev*

Cauda, F *et al* (2012) Meta-analytic clustering of the insular cortex Characterizing the meta-analytic connectivity of the insula when involved in active tasks, *Neuroimage*, 62 (1) 343–355

Damasio, A (2008) *Descartes' Error: Emotion, reason and the human brain*, Random House, London

Gasquoine, P G *et al* (2014) Contributions of the Insula to Cognition and Emotion, *Neuropsychol Rev*, 24, 77–87

Ibrahim, C *et al* (2019) The Insula: A Brain Stimulation Target for the Treatment of Addiction, *Frontiers in Pharmacology*, 10, 720

images showing painful events forms subjective pain representation in humans brain, *Cerebral Cortex*, 17 (5) 1139–46

Namkung, H *et al* (2017) The insula: an underestimated brain area in clinical neuroscience, psychiatry, and neurology, *Trends Neurosci*, 40 (4) 200–207

Naqvi, N H *et al* (2014) The insula: a critical neural substrate for craving and drug seeking under conflict and risk, *Ann N Y Acad Sci*, 1316, 53–70

Ogino, Y *et al* (2007) Inner experience of pain: imagination of pain while viewing images showing painful events forms subjective pain representation in humans brain, *Cerebral Cortex*, **17** (5), pp 1139–46

Rolls, E T (2015) Functions of the anterior insula in taste, autonomic, and related functions, *Brain and Cognition*, 110, 4–19

Uddin, L Q *et al* (2017) Structure and function of the human insula, J Clin Neurophysiol 34 (4) 300–306

Chapter 5

Andrewes, D G, Jenkins, L M (2019) The Role of the Amygdala and the Ventromedial Prefrontal Cortex in Emotional Regulation: Implications for Post–traumatic Stress Disorder, *Neuropsychology Review*

Bocchio, M *et al* (2017) Synaptic Plasticity, Engrams and Network Oscillations in Amygdala Circuits for Storage and Retrieval of Emotional Memories, *Neuron Review*, 94 (4) 731–743

Cowan, C S M *et al* (2017) Gutsy Moves: The Amygdala as a Critical Node in Microbiota to Brain Signaling, *BioEssays Journal*

Diano, M *et al* (2017) Amygdala Response to Emotional Stimuli without Awareness: Facts and Interpretations, *Frontiers in Psychology*, 7, 2029

Gangopadhyay, P *et al* (2020) Prefrontal–amygdala circuits in social decision-making, *Nature Neuroscience*, 24, 5–18

Gothard, K M (2020) Multidimensional processing in the amygdala, *Nat Rev Neurosci*, 21 (10) 565–575

Lee, J H *et al* (2016) Amygdala Circuits for Fear Memory: A Key Role for Dopamine Regulation, *The Neuroscientist*, 1–12

Chapter 6

Apps, M A J *et al* (2016) The Anterior Cingulate Gyrus and Social Cognition: Tracking the Motivation of Others, *Neuron Review*, 90 (4) 692–707

Choo, W C *et al* (2005) Dissociation of cortical regions modulated by both working memory load and sleep deprivation and by sleep deprivation alone, *Neuroimage*, **25** 579–87

Heilbronner, S R, and Hayden, B Y, (2016) Dorsal Anterior cingulate Cortex: A Bottom-Up View, *Annu Rev Neurosci 08* (39) 149–170

Molenberghs, P and Louis, W R (2018) Insights From fMRI Studies Into Ingroup Bias, *Frontiers in Psychology*

Shenhav, A *et al* (2013) The expected value of control: An integrative theory of anterior cingulate cortex function, *Neuron*, 79 (2) 217–240

Smith, R *et al* (2019) The role of anterior and midcingulate cortex in emotional awareness: A domain-general processing perspective, *Handbook of Clinical Neurology*, 166 Cingulate cortex, 6

Touroutoglou, A and Dickerson, B C (2019) Cingulate-centered large-scale networks: Normal functions, aging, and neurodegenerative disease, *Handbook of Clinical Neurology*, 166, Cingulate cortex, 8

Chapter 7

Dongsheng, C *et al* (2013) Hypothalamic programming of systemic ageing involving IKK-β, NF-κB and GnRH, *Nature Journal*, 497, 211–16

Chapter 8

Avchalumov, Y and Mandyam, C D (2021) Plasticity in the Hippocampus, Neurogenesis and Drugs of Abuse, *Brain Sciences*, 11

Bettio, L E B *et al* (2017) The Effects of Aging in the Hippocampus and Cognitive Decline, *Neurosci Biobehav Rev*, 79, 66–86

Biderman, N *et al* (2020) What Are Memories For? The Hippocampus Bridges Past Experience with Future Decisions, *Trends in Cognitive Sciences*, 24 (7) 542–556

Bryant, K G and Barker, J M (2020) Arbitration of Approach Avoidance Conflict by Ventral Hippocampus, *Frontiers in Neuroscience*

Buzsaki, G and Tingley D (2018) Space and time: The hippocampus as a sequence generator, *Trends Cogn Sci*, 22 (10) 853–869

Ekstrom, A D and Ranganath, C (2018) Space, Time and Episodic Memory: The Hippocampus is all over the Cognitive Map, *Hippocampus*, 28 (9) 680–687

Erickson, K I *et al* (2010) Exercise training increases size of hippocampus and improves memory, *Proceedings of the National Academy of Sciences of the United States of America*, 108 (7), 3017–22

Hansen, N (2017) The Longevity of Hippocampus+Dependent Memory Is Orchestrated by the Locus coeruleus-Noradrenergic System, *Neural Plast*

Hebb, D O (2002) *The Organization of Behavior: A neuropsychological theory*, Psychology Press, London (Avchalumov, 2021)

Heijer, T D *et al* (2011) A study on the bidirectional association between hippocampal volume on magnetic resonance imaging and depression in the elderly, *Biological Psychiatry*, 70, 191–97

Jurkowski, M P *et al* (2020) Beyond the Hippocampus and the SVZ: Adult Neurogenesis Throughout the Brain, *Frontiers in Cellular Neuroscience*

Kim, E J *et al* (2015) Stress effects on the hippocampus: a critical review, *Learn Mem*, 22 (9) 411–416

Lisman, J *et al* (2017) Viewpoints: how the hippocampus contributes to memory, navigation and cognition, *Nat Neurosci* 20 (11) 1434–1447

Mack, M L *et al* (2018) Building concepts one episode at a time: The hippocampus concept formation, *Neurosci Lett* 27, 680, 31–38

Maguire, E A *et al* (2006) London taxi drivers and bus drivers: a structural MRI and neuropsychological analysis, *Hippocampus*, 16, 1091–101

Weilbacher R A and Gluth, S (2016) The Interplay of Hippocampus and Ventromedial Prefrontal Cortex in Memory-Based Decision Making, *Brain Sciences*, 7 (1) 4

Chapter 9

Adam, E K *et al* (2006) Day-to-day dynamics of experience–cortisol associations in a population-based sample of older adults, *Proceedings of the National Academy of Sciences of the United States of America*, **103** (45), pp 17058–63

Amore, G *et al* (2012) A Focus on the Cerebelum: From Embryogenesis to an Age-Related Clinical Perspective, *Front Syst Neurosci*

Bares, M *et al* (2018) Consensus paper: Decoding the Contributions of the Cerebellum as a Time Machine From Neurons to Clinical Applications, *The Cerebellum* 18 (2)

Bares, M *et al* (2019) Consensus paper: Decoding the Contrirbutions of the Cerebellum as a Machine From Neurons to Clinical Applications, *The Cerebellum*, 18 (2)

Ceylan, M E at al (2015) The Contribution of the Cerebellum in the Hierarchial Development of the Self, *The Cerebellum*, 14 (6) 711–21

Diedrichsen, J *et al* (2019) Universal Transform or Multiple Functionality? Understanding the Contribution of the Human Cerebellum across Task Domains, *Neuron*, 102 (5) 918–928

Dunn, E W *et al* (2008) Spending money on others promotes happiness, *Science Magazine*, **319** (5870), pp 1687–88

Fjorback, L O *et al* (2011) Mindfulness-based stress reduction and mindfulness-based cognitive therapy: a systematic review of randomized controlled trials, *Acta Psychiatrica Scandinavica*, **124**, pp 102–19

Hull, C (2020) Prediction signals in the cerebellum: Beyond supervised motor learning, *eLife*, 9, e54073

Klaus, J and Schutter, D J L G (2021) Functional topography of anger and aggression in the human cerebellum, *NeuroImage*, 226

Lange, I *et al* (2015) The Anatomy of Fear Learning in the Cerebellum: A Systematic Meta-analysis, *Neurosci Biobehav Rev*, 59, 83–91

Leggio, M and Olivito, G (2018) Topography of the cerebellum in relation to social brain regions and emotions, *Handbook of Clinical Neurology*, 154, 5

Overwalle, F V *et al* (2015) Social Cognition and the Cerebellum: A Meta-Analytic Connectivity Analysis, *Human Brain Mapping*, 36, 5137–5154

Peterburs, J and Desmond, J E (2016) The role of the human cerebellum in performance monitoring, *Curr Opin Neurobiol*, 40, 38–44

Phillips, J R *et al* (2015) The cerebellum and psychiatric disorders, *Frontiers in Public Health*

Sokolov, A A *et al* (2017) The Cerebellum: Adaptive Prediction for Movement and Cognition, *Trends Cogn Sci*, 21 (5) 313–332

Wagner, M J and Luo, L (2020) Neuocortex-cerebellum circuits for cognitive processing, *Trends Neurosci*, 43 (1) 42–54

Part Two

Chapter 10

Adam, E K *et al* (2017) Diurnal Cortisol Slopes and Mental and Physical Health Outcomes:A Systematic Review and Meta-analysis, *Psychoneuroendocrinology* 83, 25–41

Boileau, K *et al* (2019) Ethnic differences in diurnal cortisol profiles in healthy adults: A meta-analysis, *Br J Health Psychol* 4, 806–827

Bunea, I M *et al* (2017) Early-life adversity and cortisol response to social stress: a meta-analysis, *Transl Psychiatry* 7, 1274

Clow, A and Smyth, N (2020) Salivary cortisol as a non-invasive window on the brain, *International Review of Neurobiology* 150, 1–16

Clow, A and Smyth, N (2020) Salivary cortisol as a non–invasive window on the brain, *Int Rev Neurobiol*, 150, 1–16

cognitive therapy: a systematic review of randomized controlled trials, *Acta Psychiatrica Scandinavica*, 124, 102–19

Dolan, R J *et al* (2010) Dopamine, time, and impulsivity in humans, *The Journal of Neuroscience*, 30, pp 8888–96

Dunn, E W *et al* (2008) Spending money on others promotes happiness, *Science Magazine*, 319 (5870) 1687–88

Fjorback, L O *et al* (2011) Mindfulness-based stress reduction and mindfulness-based

Law, R and Clow, A (2020) Stress, the cortisol awakening response and cognitive function, *International Review of Neurobiology* 150, 187–217

Mason, T M and Duffy, A R (2018) Complicated Grief and Cortisol Response: An Integrative Review of the Literature, *J Am Psychiatr Nurses Assoc* 25 (3) 181–188

Meyer, D and Sledge, R (2020) The potential role of cortisol as a biomarker of physiological interdependence in romantic couples: A systematic review, *Psychoneuroendocrinology* 121

O'Leary, K *et al* (2015) A systematic review of the effects of mindfulness interventions on cortisol, *J Health Psychol* 21 (9) 2108–2121

Sharot, T *et al* (2012) How dopamine enhances an optimism bias in humans, *Current Biology*, **22**, pp 1477–81

Smyth, N *et al* (2020) Effectiveness of stress-relieving strategies in regulating patterns of cortisol secretion and promoting brain health, *International Review of Neurobiology* 150, 219–246

Chapter 11

Cools, R (2016) The costs and benefits of brain dopamine for cognitive control, *WIREs Cogn Sci* 7, 317–329

Cools, R (2019) Chemistry of the Adaptive Mind: Lessons from Dopamine, *Neuron* 104, 113–131

Cools, R *et al* (2016) Dopamine and the motivation of cognitive control, *Handbook of Clinical Neurology* 163, 123–143

Dolan, R J *et al* (2010) Dopamine, time, and impulsivity in humans, *Journal of Neuroscience*, 30, 8888–96

Gebauer, L *et al* (2016) Oxytocin improves synchronisation in leader-follower interaction, *Scientific Reports*, **6**. 38416 doi:10.1038/srep38416

https://www.scientificamerican.com/article/fact-or-fiction-oxytocin-is-the-love-hormone/ (archived at https://perma.cc/APJ8-QEQM) [Online] Accessed 14 January 2017

Sharot, T *et al* (2012) How dopamine enhances an optimism bias in humans, *Current Biology*, 22, 1477–81

Steinman, M Q *et al* (2015) Sex-specific effects of stress on oxytocin neurons correspond with responses to intranasal oxytocin, *Biological Psychiatry*. doi:10.1016/j.biopsych.2015.10.007

UCI News (2015) https://news.uci.edu/research/love-hormone-helps-produce-bliss-molecules-to-boost-pleasure-of-social-interactions (archived at https://perma.cc/B39Z-Y5ZC) [Online] Accessed 14 January 2017

Chapter 12

Crespi, B J (2015) Oxytocin, testosterone, and human social cognition, *Biological Reviews*

De Dreu, C K W and Kret, M E (2015) Oxytocin Conditions Intergroup Relations Through Upregulated In–Group Empathy, Cooperation, Conformity, and Defense, *Biological Psychiatry* 79 (3) 165–173

Ebert, A and Brune, M (2017) Oxytocin and Social Cognition, *Current Topics in Behavioral Neurosciences*, 35, 375–388

Erdozain, A M and Penagarikano, O (2020) Oxytocin as Treatment for Social Cognition, Not There Yet, *Frontiers in Psychiatry* 10

Gebauer, L *et al* (2016) Oxytocin improves synchronisation in leader-follower interaction, *Scientific Reports*, 6 38416

Hurlemann, R and Marsh, N (2016) Deciphering the modulatory role of oxytocin in human altruism, *Reviews in the Neurosciences* 28 (4)

Hurlemann, R and Scheele, D (2016) Dissecting the Role of Oxytocin in the Formation and Loss of Social Relationships, *Biological Psychology* 79 (3) 185–193

Maroun, M and Wagner, S (2016) Oxytocin and Memory of Emotional Stimuli: Some Dance to Remember, Some Dance to Forget, *Biological Psychiatry* 79 (3) 203–212

Matsushita, H *et al* (2019) Oxytocin and Stress: Neural Mechanisms, Stress–Related Disorders, and Therapeutic Approaches, *Neuroscience* 417, 1–10

Neumann, I D and Slattery D A (2016) Oxytocin in General Anxiety and Social Fear: A Translational Approach, *Biological Psychology* 79 (3) 213–221

Piva, M and Chang, S W C (2018) An integrated framework for the role of oxytocin in multistage social decision–making, *American Journal of Primatology* 80 (10)

Quintana, D S and Guastella A J (2020) An Allostatic Theory of Oxytocin, *Opinion* 24 (7) 515–528

Quintana, DS *et al* (2020) Advances in the field of intranasal oxytocin research: lessons learned and future directions for clinical research, *Molecular Psychiatry* 26, 80–91

Steinman, M Q *et al* (2019) Complementary neural circuits for divergent effects of oxytocin: social approach versus social anxiety, *Biological Psychology* 85 (10) 792–801

UCI News (2015) https://newsuciedu/research/love-hormone-helps-produce-bliss molecules-to-boost-pleasure-of-social-interactions [Online]

Young, L J (2014) Oxytocin, Social Cognition and Psychiatry, *Neuropsychopharmacology*, 40 (1) 243–244

Chapter 13

Carhart–Harris, R L and Nutt, D J (2017) Serotonin and brain function: a tale of two receptors, *Journal of Psychpharmacology* 31 (9) 1091–1120

Field, T *et al* (2004) Massage therapy effects on depressed pregnant women, *Journal of Psychosomatic Obstetrics and Gynaecology*, 25, 115–22

Chapter 14

Borodovitsyna, O *et al* (2018) Persistent Stress–Induced Neuroplastic Changes in the Locus Coeruleus/Norepinephrine System, *Neural Plasticity*

McBurney–Lin, J *et al* (2019) Locus Coeruleus-Norepinephrine Modulation of Sensory Processing and Perception: A Focused Review, *Neurosci Biobehav Rev* 105, 190–199

Ross J A and Van Bockstaele E J (2021) The Locus Coeruleus– Norepinephrine System in Stress and Arousal: Unraveling Historical, Current, and Future Perspectives, *Frontiers in Psychiatry* 11, 1581

Unsworth, N and Robison, M K (2017) A locus coeruleus-norepinephrine account of individual differences in working memory capacity and attention control, *Psychon Bull Rev* (4) 1282–1311

Wahis, J and Holt, M G (2021) Astrocytes, Noradrenaline, α1–Adrenoreceptors, and Neuromodulation: Evidence and Unanswered Questions, *Frontiers in Cellular Neuroscience* 15,

Xing, B *et al* (2016) Norepinephrine versus Dopamine and their Interaction in Modulating Synaptic Function in the Prefrontal Cortex, *Brain Res* 1641 (B) 217–233

Part Three

Chapter 16

Brann, A (2013) *Make Your Brain Work: How to maximize your efficiency, productivity and effectiveness*, Kogan Page, London

Draganski, B *et al* (2006) Temporal and spatial dynamics of brain structure changes during extensive learning, *The Journal of Neurosience*, 26 (23), pp 6314–17

Hebb, D O (2002) *The Organization of Behavior: A neuropsychological theory*, Psychology Press, London

Li, S-C *et al* (2008) Working memory plasticity in old age: practice gain, transfer, and maintenance, *Ulman Psychology and Aging*, 23, pp 731–42

Chapter 17

Dweck, C S (2007) *Mindset: The New Psychology of Success*, Ballantine Books

Chapter 19

Ericsson, K A (1980) Acquisition of a memory skill, *Science*, **208**, pp 1181–82
Hirano, Y *et al* (2008) Effects of chewing in working memory processing, *Neuroscience Letters*, **436** (2), pp 189–92
Miller, G A (1956) The magical number seven, plus or minus two: some limits on our capacity for processing information, *Psychological Review*, **63** (2), pp 81–97

Chapter 20

Langelaan, S *et al* (2006) Do burned-out and work-engaged employees differ in the functioning of the hypothalamic-pituitary-adrenal axis?, *Scandinavian Journal of Work, Environment and Health*, **32**, pp 339–48

Chapter 21

Keysers, C (2011) *The Empathic Brain*, Social Brain Press, London
Keysers, C *et al* (2003) Both of us disgusted in my insula: the common neural basis of seeing and feeling disgust, *Neuron*, **40**, pp 655–64
New York Times (2006) [accessed 7 January 2014] Cells That Read Minds [Online] http://www.nytimes.com/2006/01/10/science/10mirr.html?pagewanted=all&_r=0 (archived at https://perma.cc/NA9L-SGCV)
Nummenmaa, L *et al* (2014) Mental action simulation synchronizes action–observation circuits across individuals, *Journal of Neuroscience*, **34**, pp 748–57

Chapter 22

McClure, S M *et al* (2004) Separate neural systems value immediate and delayed monetary rewards, *Science*, **306** (5695), pp 503–07

Part Four

Chapter 23

Bohr, N (1995) *The Philosophical Writings of Niels Bohr, Vol. 3: Essays 1958–1962 on atomic physics and human knowledge*, Ox Bow Press, Oxford

Hameroff, S and Penrose, R (2013a) Consciousness in the universe: A review of the 'Orch OR' theory, *Physics of Life Reviews*. doi:10.1016/j.plrev.2013.08.002

Hameroff, S and Penrose, R (2013b) Reply to criticism of the 'Orch OR qubit' – 'Orchestrated objective reduction' is scientifically justified, *Physics of Life Reviews*. doi:10.1016/j.plrev.2013.11.00

McTaggart, L (2012) *The Field: The quest for the secret force of the universe*, HarperCollins, London

Ochsner, K N (2002) Rethinking feelings: an FMRI study of the cognitive regulation of emotion, *Journal of Cognitive Neuroscience*, **14**, pp 1215–29

Pashler, H E (1999) *The Psychology of Attention*, MIT Press, Cambridge MA

Schwartz, J M, Stapp, H P and Beauregard, M (2005) Quantum physics in neuroscience and psychology: a neurophysical model of mind–brain interaction, *Philosophical Transactions of the Royal Society*, **360**, pp 1309–27

Stapp, H P (2011) *Mindful Universe: Quantum mechanics and the participating observer*, Springer, New York

Part Five

Chapter 24

Baumeister, R F *et al* (1998) Ego depletion: is the active self a limited resource?, *Journal of Personality and Social Psychology*, **74** (5), pp 1252–65

Casey, B J *et al* (2011) Behavioral and neural correlates of delay of gratification 40 years later, *Proceedings of the National Academy of Sciences of the United States of America*, **108** (36), pp 14998–15003

Dweck, C S *et al* (2013) [accessed 7 May 2014] Beliefs about willpower determine the impact of glucose on self-control [Online] http://www.pnas.org (archived at https://perma.cc/G54F-EH4U)

Mischel, W *et al* (1970) Attention in delay of gratification, *Journal of Personality and Social Psychology*, **16** (2), pp 329–37

Part Six

Chapter 25

Crews, F T *et al* (2009) Impulsivity, frontal lobes and risk for addiction, *Pharmacology, Biochemistry and Behaviour*, **93** (3), pp 237–47

Habib, R *et al* (2010) Neurobehavioral evidence for the 'near-miss' effect in pathological gamblers, *Journal of the Experimental Analysis of Behaviour*, **93** (3), pp 313–28

Hebb, D O (2002) *The Organization of Behavior: A neuropsychological theory*, Psychology Press, London

Smith, K S *et al* (2012) Reversible online control of habitual behavior by optogenetic perturbation of medial prefrontal cortex, *Proceedings of the National Academy of Sciences of the United States of America*, **109** (46), pp 18932–37

Chapter 26

Lowenstein, G (1987) Anticipation and the valuation of delayed consumption, *The Economic Journal*, **97** (387), pp 666–84

MacLeod, C *et al* (2002) Induced processing biases have causal effects on anxiety, *Cognition and Emotion*, **16** (3), pp 331–54

Segerstrom, S (2007) Optimism and immunity: do positive thoughts always lead to positive effects? *Brain, Behaviour and Immunity*, **19** (3), pp 195–200

Sharot, T (2012) *The Optimism Bias: Why we're wired to look on the bright side*, Robinson, London

Chapter 27

Berkman, E T, Falk, E B and Lieberman, M D (2012) [accessed 6 May 2014] Effects of Three Core Goal Pursuit Processes on Brain Control Systems: Goal Maintenance, Performance Monitoring, and Response Inhibition [Online] http://plosone.org (archived at https://perma.cc/PMH3-5ZLT)

Gollwitzer, P and Wieber, F (2010) Overcoming procrastination through planning, in *The Thief of Time: Philosophical essays on procrastination*, ed C Andreou and M D White, Oxford University Press, Oxford

Johnson, M K *et al* (2006) Dissociating medial frontal and posterior cingulate activity during self-reflection, *Social Cognitive and Affective Neuroscience*, **1**, pp 56–64

Lieberman, M D *et al* (2007) Putting feelings into words: affect labelling disrupts amygdala activity in response to affective stimuli, *Psychological Science*, **18** (5), pp 421–28

Spunt, R P and Lieberman, M D (2014) Automaticity, control, and the social brain, in *Dual Process Theories of the Social Mind*, ed J Sherman, B Gawronski and Y Trope, The Guilford Press, New York

Chapter 28

Brown, K W and Ryan, R M (2003) The benefits of being present: mindfulness and its role in psychological well-being, *Journal of Personality and Social Psychology*, **84**, pp 822–48

Dickenson, J *et al* (2013) Neural correlates of focused attention during a brief mindfulness induction, *Social Cognitive and Affective Neuroscience*, **8**, pp 40–47

Jain, S *et al* (2007) A randomized controlled trial of mindfulness meditation versus relaxation training: effects on distress, positive states of mind, rumination, and distraction, *Annals of Behavioural Medicine*, **33**, pp 11–21

Jha, A P *et al* (2010) Examining the protect effects of mindfulness training on working memory capacity and effective experience, *Emotion*, **10**, pp 54–64

Lazar, S *et al* (2005) Meditation experience is associated with increased cortical thickness, *Neuro-Report*, **16**, pp 1893–97

Lutz *et al* (2008) Regulation of the neural circuitry of emotion by compassion meditation: effects of meditative expertise, *PLoS ONE*, **3** (3), pp 1–10

Tang, Y Y and Posner, M I (2009) Attention training and attention state training, *Trends in Cognitive Sciences*, **13**, pp 222–27

Tang, Y Y *et al* (2010) Short-term meditation induces white matter changes in the anterior cingulate, *Proceedings of the National Academy of Sciences of the United States of America*, **107**, pp 15649–52

Chapter 29

Pates, J, Cummings, A and Maynard, I (2002) The effects of hypnosis on flow states and three-point shooting performance in basketball players, *The Sport Psychologist*, **16**, pp 1–15

Sachs, O and Hirsch, J (2008) A neurology of belief, Annals of Neurology, 63, 129–130

Science Direct. www.sciencedirect.com/science/article/pii/S0278262619303860#b0450 (archived at https://perma.cc/VC9T-CKTH)

Chapter 30

Glucksberg, S (1962) The influence of strength of drive on functional fixedness and perceptual recognition, *Journal of Experimental Psychology*, **63**, pp 36–41

Pink, D H (2011) *Drive*, Canongate Books Ltd, Edinburgh

Chapter 31

Cikara, M *et al* (2013) Their pain, our pleasure: stereotype content and schadenfreude, *Annals of the New York Academy of Sciences*, **1299**, pp 52–59

Damasio, A R (1996) The somatic marker hypothesis and the possible functions of the prefrontal cortex, *Philosophical Transactions of the Royal Society of London*, **351**(1346), pp 1413–20

Sanfey, A G (2007) Social decision-making: insights from game theory and neuroscience, *Science*, **318** (5850), pp 598–602

Chapter 32

Scott, D J *et al* (2007) Individual differences in reward responding explain placebo-induced expectations and effects, *Neuron*, **55**, pp 325–36

Volkow, N D *et al* (2004) The addicted human brain viewed in the light of imaging studies: brain circuits and treatment strategies, *Neuropharmacology*, **47**, pp 3–13

Chapter 33

Damasio, A R (2000) Thinking about belief, in *Memory, Brain and Belief*, ed D L Schacter and E Scarry, Harvard University Press, Cambridge MA

Lieberman, M D *et al* (2005) An fMRI investigation of race-related amygdala activity in African-American and Caucasian-American individuals, *Nature Neuroscience*, **8**, pp 720–22

Psychiatric Association Handbook (1994) American Psychiatric Association, Arlington VA

Taylor, K (2006) *Brainwashing: The science of thought control*, Oxford University Press, Oxford

Winston, J S *et al* (2002) Automatic and intentional brain responses during evaluation of trustworthiness of faces, *Nature Neuroscience*, **5**, pp 277–83

Chapter 34

Arch, J *et al* (2012) Randomized clinical trial of cognitive behavioral therapy (CBT) versus acceptance and commitment therapy (ACT) for mixed anxiety disorders, *Journal of Consulting and Clinical Psychology*, **80** (5), October, pp 750–765 [Online] http://dx.doi.org/10.1037/a0028310 (archived at https://perma.cc/TPZ6-SX5V)

Bargh, J A *et al* (1996) Automaticity of social behavior: direct effects of trait construct and stereotype activation on action, *Journal of Personality and Social Psychology*, **71**, pp 230–44

Garner, R (2005) Post-it note persuasion: a sticky influence, *Journal of Consumer Psychology*, **15** (3) pp 230–37

Hausman, D M and Welch, B (2009) Debate: to nudge or not to nudge, *Journal of Political Philosophy*, **18** (1), pp 124–36

Chapter 35

Daybreak's Stranger Danger, ITV, 4–5 September 2013

Loftus, E F *et al* (2002) Make my memory: how advertising can change our memories of the past, *Psychology & Marketing*, **19** (1), pp 1–23

Loftus, E F *et al* (2013) Misinformation can influence memory for recently experienced, highly stressful events, *International Journal of Law and Psychiatry*, **36**, pp 11–17

Patihis, L *et al* (2013) False memories in highly superior autobiographical memory individuals, *Proceedings of the National Academy of Sciences*, **110**, 20947–52, doi:10.1073/pnas.1314373110

Part Seven

Chapter 36

Covey, S [accessed 6 May 2014] How the Best Leaders Build Trust [Online] http://leadershipnow.com (archived at https://perma.cc/88AG-8E63)

Zak, P (2012) *The Moral Molecule: How trust works*, Penguin Group US, New York

Chapter 37

Singer, T *et al* (2006) Empathic neural responses are modulated by the perceived fairness of others, *Nature*, **439**, pp 466–69

Tabibnia, G and Lieberman, M D (2007) Fairness and cooperation are rewarding: evidence from social cognitive neuroscience, *Annals of the New York Academy of Science*, **18**, pp 90–101

Chapter 38

Baumeister, R F *et al* (2005) Social exclusion impairs self-regulation, *Journal of Personality and Social Psychology*, **88**, pp 589–604

Cacioppo, J T (2009) *Loneliness: Human nature and the need for social connection*, W W Norton & Company, New York

Cacioppo, J T *et al* (2000) Lonely traits and concomitant physiological processes: the MacArthur social neuroscience studies, *International Journal of Psychophysiology*, **35**, pp 143–54

Cacioppo, J T *et al* (2009) In the eye of the beholder: individual differences in loneliness predict neural responses to social stimuli, *Journal of Cognitive Neuroscience*, **21**, pp 83–92

Kanai, R *et al* (2012) Brain structure links loneliness to social perception, *Current Biology*, **22**, pp 1975–79

Lieberman, M D *et al* (2003) Does rejection hurt? An fMRI study of social exclusion, *Science Magazine*, **302**, pp 290–92

INDEX

CPSIA information can be obtained
at www.ICGtesting.com
Printed in the USA
BVHW020319020422
632997BV00005BA/35